EXPERIMENTS IN ASTRONOMY

Third Edition

Leo Blitz

Michael F. A' Hearn

University of Maryland
College Park, Maryland

PEARSON

Custom
Publishing

Copyright © 1991, 1989, 1983 by Leo Blitz
Bellwether Press Division
ISBN 0-8087-7393-3

Printed in the United States of America.
30   29   28   27   26   25   24   23   22   21

PEARSON CUSTOM PUBLISHING
75 Arlington Street, Suite 300, Boston, MA 02116
A Pearson Education Company

Preface to First Edition

This book has been written for the laboratory portion of the introductory astronomy course for non-science majors of the University of Maryland. The aim of the book is to provide quantitative exercises which illustrate a number of the major concepts taught in the one semester course at the university. Most of the labs are designed to be completed in the two hour laboratory sessions. A number of the labs use simple equipment or computer programs which must be manipulated by the students; most of the rest require measurements to be made from high quality reproductions of astronomical photographs. Since some of our students take the lab source one or more semesters after taking the lecture course, the labs have a substantial amount of introductory material. The introductory sections are also meant to minimize the "cookbook" aspects of the course which are hard to avoid in a lab course for non-science majors.

This edition of the lab book represents an evolution from the previous versions. Some of the previous labs used in this edition have only small modifications, some have been extensively rewritten, and four are completely new. As such, it is important to acknowledge the people who have contributed significantly to previous versions of this book and who have developed the original versions of many of the labs that have been adopted in this new edition. These include Don Wentzel, Mary Spanburgh, George Dunkelberger, Elske Smith, Tom Matthews, Mike A'Hearn, Herb Frey, Dave Schleicher and Dave Theison. Don Wentzel also provided help finding and correcting errors. I have been helped at various junctures by John Trasco, Eric Smith and Loris Magnani. Jan Hall and Betty Stevenson provided, fast, efficient and accurate typing of the manuscript. Joan Wright drafted many of the illustrations.

Although the lab book is really the culmination of the efforts of many people extending over along time, it is of course I who am responsible for any errors and deficiencies in this version.

Leo Blitz
College Park
June 1986

Preface to Second Edition

The exercises used in this course undergo continual evolution as new people teach the course. Since the previous edition of this manual was written, a number of changes have been introduced, often in response to suggestions from the many instructors, both graduate students and faculty, who have taught the exercises in the classroom. Additional changes have been introduced to incorporate developments in technology, specifically the proliferation of computers, or to change the emphasis in some exercises.

The atlas of standard stars as well as the spectra of unknown stars used in *Stellar Spectra* have been replaced with theoretical, calculated spectra, ©Roger A. Bell, produced by Roger Bell and Michael Briley. These synthetic spectra, which have been quantitatively compared with real stellar spectra, remove the vagaries of variable exposure times and variable reddening that caused many students confusion in the earlier versions of the exercise.

The exercises of *Galactic Rotation* and *Expansion of the Universe* have been totally rewritten so that they are executed on a personal computer. The computer programs for these labs, as well as improved and updated versions of the programs previously used in *Jupiter's Moons* and *Parallax* have been developed by Chad McDaniel and Jon Greenblatt of the Academic Software Development Group of the University of Maryland in collaboration with myself and John Trasco who is leading the Astronomy Program's participation in our campus-wide program to expand the use of personal computers in courses. The four rewritten labs and the computer software are ©University of Maryland and are reprinted here by permission. The computer programs for these labs are distributed by:

Academic Software Development Group
University of Maryland
College Park, MD 20742
(301) 405-7600

I would like to thank the many individuals who taught these labs over the last couple of years and made suggestions. I would also like to thank Mike Briley for assistance in making graphs and text for Lab 7, Doug Gill for interesting discussions of fireflies which were stimulated by a suggestion from Paul Butler, and Maggie Berry for preparing the manuscript, both the typing and the layout.

Michael F. A'Hearn
College Park
August 1990

Preface to Third Edition

This edition corrects many typos, phraseology, and ambiguities from the previous edition and includes one additional lab, on solar activity, written by Don Wentzel, reprinted with his permission and based on exercises previously developed by himself and Elske Smith. The software used in *Galactic Rotation* and *Expansion of the Universe* utilizes previous software developments by R. A. Bell and G. Beam, an acknowledgement which I failed to make previously. I thank particularly M. Berry, T. Helfer, L. Mundy, and M. Thornley for extensive comments on the previous edition.

Michael F. A'Hearn
College Park
August 1991

Introduction

The purpose of this laboratory course is to give you some insight into how astronomers take and use data to reach conclusions about the cosmos. What you learn will, however, have much broader applicability. Everyone at some time in his or her life must make use of numerical data in order to reach conclusions or to make decisions. For example, if you are running a business and you see that your sales are changing from month to month in a well defined way, you need to know how to use your sales information to increase or decrease your inventory. Or if you are in charge of a food distribution program in some area recently hit by disaster, you need to know how to estimate how many people will be in need of relief so that you can request the right number of supplies. You need to know not only how to use data that are supplied to you, but how to determine what data you need and how to acquire them.

In this course you will learn certain skills in the acquisition, handling and interpretation of data. You will learn how to analyze the information you have obtained in order to reach sensible conclusions. The skills you acquire will have general applicability to many different situations outside of astronomy.

This is, however, an astronomy course. You will be obtaining, using, and interpreting data in an astronomical context. Each exercise is designed to show you how astronomers use a particular set of data to learn about a particular facet of astronomy. In this manual the exercises begin in the classroom, move to Earth and sky, to the solar system, to stars and the interstellar medium, to galaxies, and finally to the expansion of the universe. At most of these stages, one of the key questions that we will ask is "How far away is this object and how do we know that?" since the astronomical distance scale is built like a pyramid with the distance to any given object relying on our knowledge of distances to many closer objects. Remember as you work through the exercises that the goal is to understand why astronomers work in the way that they do, not just to get a correct numerical answer.

Check that your manual includes an envelope with 6 large photographs. Keep these in a safe place until you need them. Four of the photographs will be used in Lab 11 and the other two will be used in Lab 13. Even the greatest scientists make errors and probably you will also. Answer questions in ink and if you make an error neatly cross it out. Graphs and markings on the copies of photographs in this manual should be done in pencil since in these two instances it is necessary to erase mistakes rather than crossing them out.

Unless indicated otherwise, the labs are to be done during one class period, normally two hours long, and submitted at the end of that class period. As in most laboratory situations, lab partners will often share data that they have taken together but each student must individually and independently answer all the questions in the lab. To be sure that you have enough time to finish the lab, it is important that you read the exercise before coming to class. The lab manual presupposes that you are taking or have already taken a lecture course in elementary astronomy and are therefore familiar with the material before beginning the lab. Thus you should be able to read the manual in a half hour or less and understand most of it before coming to class. If you do not read this material before coming to class, you are likely to be bewildered in class and not finish the lab in the allotted time.

Although our primary goal is to develop an understanding of scientific methods, we hope that you will also enjoy doing these exercises. Most astronomers "do" astronomy because they find it fun and that is certainly the reason why many students take courses in astronomy. If you get too bogged down in the details, you will likely not enjoy the course so try to take the time to step back and ask yourself where each exercise fits in the scale of the universe. If you develop that perspective you will also be able to enjoy the work.

Contents

1

Name _____

Section _____

Lab Partner _____

Mathematical Tools

PURPOSE: To review all the arithmetic, algebraic, and graphing tools that you need to complete the labs in this course.

EQUIPMENT: A straight edge for each student. A two-meter stick and a bathroom scale for the entire class.

REQUIREMENTS: The problems on pages 1 through 12 are to be done individually. The measurements for the remainder of the lab will be done by the class as a group. The questions and graphs for the remainder of the lab are to be done individually.

REMEMBER: Nothing you do in the rest of the labs will involve more advanced mathematical tools than the ones you will use in this lab. Do the lab with care.

INTRODUCTION

The material in this course is basically quantitative. Astronomers must always ask, "How much?" "How far? How big?", and they must routinely work with the numbers they get from observations. The level of mathematics required to do the laboratory material is no greater than what is required for admission to the university. Nevertheless, to make sure that everyone is on an equal footing you will be required to complete this lab before you begin the next one. The basic math you will review is the following: simple algebraic manipulation, scientific notation, graphing, and manipulation of units. Each of these is covered in turn below.

ALGEBRA

The most complex equations you will have to use have the form:

$$ax + b = y$$

If the values of a, b and y are known, the value of x can then be determined. Given an equation of this form, you should know how to solve for x. The general rule is: as long as you do the same thing to both sides of an equation, you leave the equation unchanged. In this case, first subtract b from both sides of the equation:

$$
\begin{array}{rcl}
ax + b & = & y \\
\underline{ - b} & = & \underline{- b} \\
ax & = & y - b
\end{array}
\qquad \text{e.g.} \qquad
\begin{array}{rcl}
3x + 6 & = & 15 \\
\underline{ - 6} & = & \underline{- 6} \\
3x & = & 9
\end{array}
$$

1

Then divide both sides by a:

$$\frac{ax}{a} = \frac{y-b}{a} \qquad\qquad \frac{3x}{3} = \frac{9}{3}$$

$$x = \frac{y-b}{a} \qquad\qquad x = 3$$

Consider, for example, an equation of the form

$$x = \frac{y}{z}.$$

What if z is a fraction, for example 1/10? That is,

$$x = \frac{y}{1/10}.$$

To simplify this equation, one must remember that multiplying any number by 1 leaves the number unchanged. Therefore, multiply the right hand side by 10/10 = 1. In that case, you have

$$x = \frac{y}{1/10} \times \frac{10}{10}$$

Since 1/10 x 10 = 1, the equation reduces to

$$x = 10\ y.$$

PROBLEMS

1. If 5x + 15 = y, what is x? _____

2. If xy = 1/5, what is y? _____

3. If 3x + 27 = y, what is x? _____

   What is x if y = 36? _____

   If y becomes 30, does x become bigger or smaller? _____

4. The velocity (v) of an object is related to the time (t) the object travels and the distance (d) it travels according to:

$$vt = d.$$

How fast, in kilometers per second, is a star moving that has traveled 1000 kilometers (km) in 40 seconds? _____

5. Consider the equation

$$T = \frac{Q}{P}$$

If the value of Q is halved and T remains constant, how does the value of P change? _____

NOTES ON NOTATION

Exponents:

$$a \times a = a^2$$

$$a \times a \times a = a^3$$

In general,

$$\underbrace{a \times a \times a...a}_{n \text{ times}} = a^n$$

Multiplication:

$$a^2 \times a^3 = (a \times a) \times (a \times a \times a) = a^5$$

$$a^4 \times a^2 = (a \times a \times a \times a) \times (a \times a) = a^6$$

In general, $a^n \times a^m = a^{(n+m)}$

3

Division:

$$\frac{a^4}{a^2} = \frac{a \times a \times a \times a}{a \times a} = a \times a = a^2$$

$$\frac{a^5}{a^3} = \frac{a \times a \times a \times a \times a}{a \times a \times a} = a \times a = a^2$$

In general,

$$\frac{a^m}{a^n} = a^{m-n} \qquad \text{e.g.} \qquad \frac{10^8}{10^5} = 10^{8-5} = 10^3$$

If n > m, then the exponent on the right hand side above is negative. What does this mean? Consider the number

$$\frac{10^3}{10^5} = \frac{1000}{100,000} = \frac{1}{100} = \frac{1}{10^2} = 10^{3-5} = 10^{-2}$$

$10^3 \div 10^5$ is a number less than one, that is, a fraction. In general, numbers with negative exponents are fractions between 0 and 1.

For $\frac{a^m}{a^n}$, if n = m this is equal to:

$$\frac{a^m}{a^m} = a^{m-m} = a^0.$$

But $\frac{a^m}{a^m} = 1$ no matter what a is.

Therefore $a^0 = 1$ for any value of a.

SCIENTIFIC NOTATION

Consider the number 376. This number can be written as 3.76 × 100 = 3.76 × (10 × 10) = 3.76 × 10^2. This way of writing numbers is called scientific notation, sometimes also known as powers-of-ten notation. In this notation, the number of "significant digits" is the number of digits written, ignoring any leading zeros. Scientists normally quote only as many significant digits as are needed for a given problem or as many as are reliably known.

Many of the numbers we deal with in astronomy are either very large or very small. For example, the number of centimeters in the

4

astronomical unit of distance called the parsec is approximately 3,090,000,000,000,000,000. Each place we count to the left from the right hand end of the number represents multiplication by 10. Thus, to write the above number in scientific notation, we write 3.09×10^{18}.

Now consider the number .000524. This is the same as

$$\frac{5.24}{10 \times 10 \times 10 \times 10} = \frac{5.24}{10^4} = 5.24 \times \left(\frac{1}{10^4}\right)$$

But $\frac{1}{10^4}$ can be written as $\frac{10^0}{10^4} = 10^{(0-4)} = 10^{-4}$. Therefore, .000524

can be written as 5.24×10^{-4}. This means that very small numbers can also be written in scientific notation. In this case we count from the decimal point to the <u>right</u> until we get past the first non-zero digit. The number of places to the right that you count is the number of times a number has been <u>divided</u> by 10, and is equal to the negative exponent in the power of 10 by which the number (in this case 5.24) must be multiplied.

Consider the following multiplication: $(3 \times 10^3) \times (4 \times 10^2)$. This is the same as 3000×400. Since multiplying numbers in any order leaves the result unchanged, we may write this as $(3 \times 4) \times (10^3 \times 10^2)$.

$$(3 \times 4) \times (10^3 \times 10^2) = (12) \times (10^5).$$

But $12 = 1.2 \times 10^1$.

$$12 \times 10^5 = 1.2 \times 10^1 \times 10^5 = 1.2 \times 10^6$$

So, in its simplest form,

$$(3 \times 10^3) \times (4 \times 10^2) = 1.2 \times 10^6.$$

PROBLEMS (Do not use a calculator.)

6. $10^5 \times 10^{10} =$ _____

7. $10^8 \times 10^4 =$ _____

8. $10^5 \div 10^{10} =$ _____

9. $10^7 \div 10^{28} =$ _____

10. $10^8 \div 10^4 =$ _____

5

Write the answers to the following in the simplest possible form with 3 significant digits, i.e. in scientific notation. (You will commonly quote answers to 3 significant digits in subsequent labs.) Calculators may be used. The instructor will show you how to keep track of exponents if you are using a simple calculator without scientific notation.

11. $(24807) \times (5.16 \times 10^3) =$ _____

12. $(4.1463 \times 10^5) \times (.603) =$ _____

13. $(3 \times 10^{16})^2 =$ _____

14. $(.0005432) \div (1287) =$ _____

15. $(5.246 \times 10^{-6}) \div (3.41 \times 10^{-6}) =$ _____

For the following, all answers need to be in scientific notation:

16. $10^2 + 10^2 =$ _____

17. $10^2 + 10^1 =$ _____

18. $(4 \times 10^8) + (5 \times 10^9) =$ _____

19. $(3.25 \times 10^{-2}) + (0.325) =$ _____

20.   Average the following numbers by first writing each number in scientific notation so that <u>all</u> numbers have the same exponent, then average.

   $2 \times 10^7 =$ _____

   10 million = _____

   half of a billion = _____

   43,000,000 = _____

   27 million = _____

   Average: _____

UNITS

   In astronomy, as in most physical sciences, most of the numbers we deal with are not pure numbers, but have units attached. For example, the distance to α Centauri, the nearest star (other than the Sun), is 1.3 <u>parsecs</u>. It is also equal to 4.1×10^{13} <u>kilometers</u>. The answers to the questions in the labs will almost always require units, and an answer without them is incomplete. For example, for the distance to α Centauri, it is not correct to say that it is 1.3 or 4.1×10^{13}; the answer depends on the units. Two rules of thumb are important.

(A)   Like apples and oranges, you cannot add two numbers together unless they have the same units.

(B)   One can multiply and divide quantities expressed in different units, but the units also get multiplied and divided.

As an example of (A), if the distance from Earth to α Centauri is 1.3 parsecs, and the distance from α Centauri to another star in the same direction is 2×10^{13} kilometers, to get the distance from the earth to the other star one must first convert parsecs to kilometers or vice versa. To do this, one uses rule B.

There are 3.09×10^{13} km (kilometers) per parsec (pc).   This may be written as

$$3.09 \times 10^{13} \ \frac{km}{pc}$$

To convert 1.3 pcs to km take

$$1.3 \ pc \times 3.09 \times 10^{13} \ \frac{km}{pc}$$

multiply: $1.3 \times 3.09 \times 10^{13} = 4.0 \times 10^{13}$ and multiply: $pc \times \dfrac{km}{pc} = km$

Therefore $1.3 \ pc = 4.0 \times 10^{13}$ km.
Consider now the equation: velocity $\times$ time = distance. If velocity = v, time = t and distance = d we have $vt = d$

We may also write this as $v = \dfrac{d}{t}$

If distance is measured in km, and time is measured in seconds, the unit of velocity becomes km/sec.

That is, if a star travels 10 km in 2 sec, then its velocity v is

$$v = \frac{10 \ km}{2 \ sec} = \frac{km}{sec}$$

The units are read as "km per second." If the distance is measured in miles and the time in hours, then the velocity is measured in $\underline{miles}$ or miles per hour.   hours

PROBLEMS

21.   A star's velocity is 30 $\frac{km}{sec}$. How far does the star
travel in 60 seconds.   _____

22.   How many seconds are there in a year?   _____

23. How far does the star in question 21 travel in 3 years? _____

24. Earth travels around the sun at approximately 30 $\frac{km}{sec}$. What is its velocity in $\frac{km}{minute}$? _____

25. Area can be expressed as (length)^2. If a square has one side equal to 5 cm (centimeters), what is its area? _____

GRAPHING

A graph in this course will always be used to represent the relationship between two sets of numbers. Consider, for example, the equation

$$5 x = y.$$

If we tabulate the values of y corresponding to the values of x for the numbers 1 through 10, we get the following:

x	y
1	5
2	10
3	15
4	20
5	25
6	30
7	35
8	40
9	45
10	50

This is a table of pairs of numbers e.g. (1, 5), (2, 10), (3, 15), etc.

The relationship between x and y can be visualized by means of a graph. (The actual graph is shown on page 12.) Each pair of numbers is represented by the distance from two perpendicular (i.e. at right angles to each other) axes which are lines signifying x and y. The point where these lines intersect is called the origin and has the value (0, 0). For example:

8

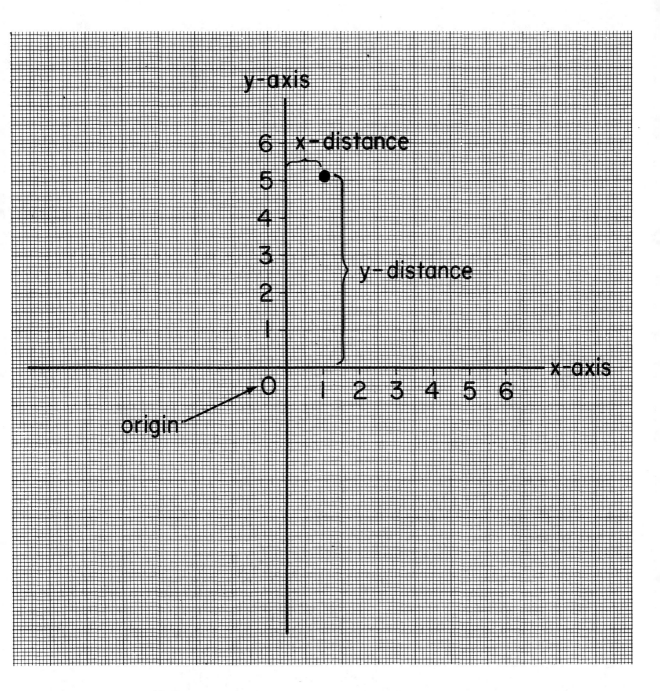

To locate a point on the graph, we count along the two axes. For example, the first point is (1, 5). Therefore count 1 unit along the x axis and 5 units along the y axis and place a clearly visible mark at the proper intersection of the two values. If we plot all the values in the table, the points look like the graph on page 12.

We can see that the points lie on a straight line. We may therefore connect the points with a straight line as shown. This line enables us to visualize the relationship between the numbers in the relationship 5x = y and to find values from the graph of number-pairs other than those in the table.

But what if the values don't lie on a straight line? Consider the equation x = 1/y. A representative table of values is the following:

x	y
1/4 = .25	4
1/3 ~ .33	3
1/2 = .5	2
1	1
2	.5
3	~ .33
4	.25

A graph of this equation is shown below. In this case we drew a smooth curve through the sets of points. The curve passed through each point because there was a simple mathematical relationship describing all the points.

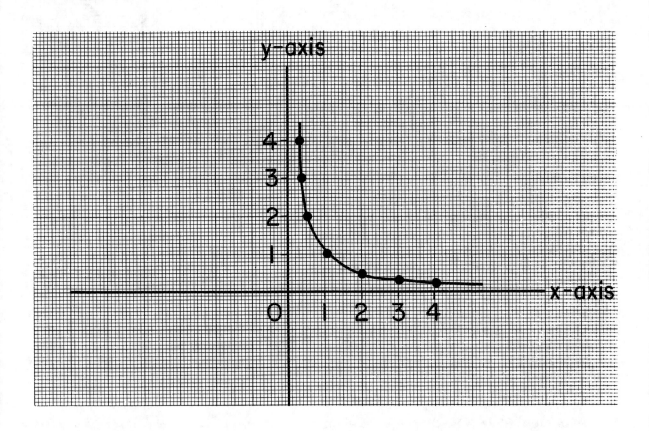

In the following example you will again consider a simple equation relating x and y and you will plot the appropriate graph.

PROBLEM

On the sheet of graph paper on the next page, make a graph of the equation

$$100 \; x = y^2$$

for integral values of x = from 0 through 10.

x	y
0.0	
1.0	
2.0	
3.0	
4.0	
5.0	
6.0	
7.0	
8.0	
9.0	
10.0	

Draw a smooth curve through all the points.

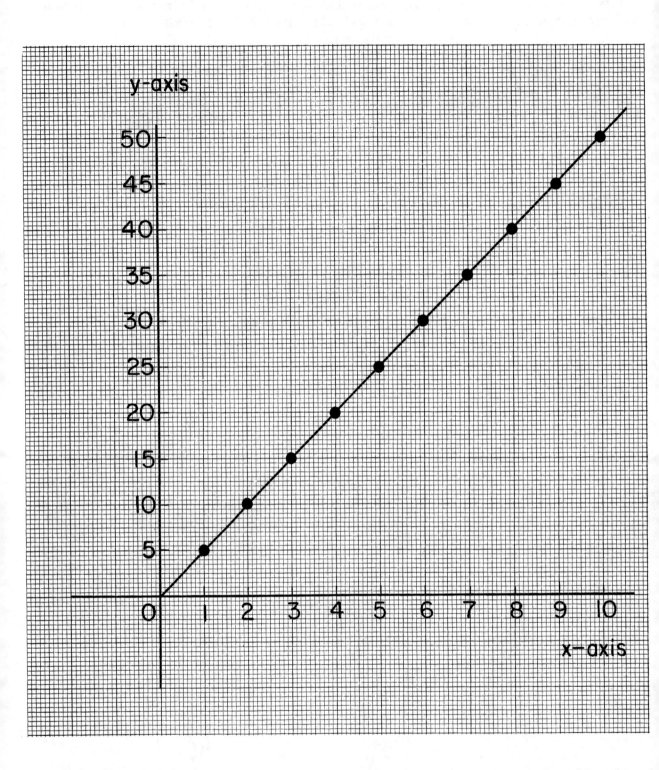

Notice that the straight line and the curve connect all the appropriate points. If the values on this graph had been based on measurement, each value would have some measurement error and the points would not all be connected even though the same line or curve represented the data.

GRAPHING OF MEASURED DATA

In this course you will be producing tables of values, like those for x and y, from measurements that you make. The purpose of drawing a graph is to see the relationship between the measured quantitites, such as between brightness and distance, between velocity and distance, and so on. When dealing with measurements on a graph, there are two factors which lead to a fundamental difference in the appearance of the graph compared to the graphs on the previous page. In the previous graphs, it was possible to draw a smooth curve, or even a straight line, which passed exactly through all the points because the points were all defined by a relatively simple equation. On graphs of real data, a smooth curve will usually not pass through all the points - points will lie on both sides of any simple, smooth curve. The two factors that cause this are intrinsic dispersion and measurement uncertainties. Intrinsic dispersion is due to the fact that the two variables are not related only by a simple equation - the relationship between "x" and "y" can also involve another variable "z" which has not been considered. Measurement uncertainty is due to the limited accuracy with which we can make any measurements.

EXERCISE

A. Measurements

Using the 2-meter stick, measure the height in centimeters for each student in the lab. Using the scale, measure the weight in pounds of each student in the lab. Enter the data in the table on page 15 being sure to put the weight for each student in the same entry as that student's height.

How large an uncertainty would you estimate that there is in the measured height for any student - 1 mm, 1 cm, or 10 cm?_____
Explain.

How large an uncertainty in the weight - 1 pound, 3 pounds, or 10 pounds?   _____
Explain.

Describe the effects that might lead you to systematically (rather than randomly) overestimate or underestimate the height or weight.

B. Finding a Relationship

Plot each student's height and weight on the graph on page 16. Be careful in plotting the points. After you have plotted all the points, you should be able to see an overall trend of weight with height. Draw a smooth line, a curved one if necessary, through the measured points that best represents the overall trend in the data. This line should not pass through every point.

Does a straight line represent the trend in the measured points adequately or does a smooth, curved line represent it better?_____

This smooth line represents the trend of weight with height for students in this lab section. Do you think this relationship would be valid for all college students? _____

For all adult human beings? _____

Explain why or why not.

How far are typical measured weights from your smooth curve? ±_____

Do you think that this scatter of points around the line is due to measurement uncertainty alone? Explain.

What would you estimate as the intrinsic scatter in the relationship? ±_____

The intrinsic scatter should, very approximately, be the typical scatter in the measured points minus the measurement uncertainty.

C. Extrapolation

Look at the smooth line through your data and using no other information extend (extrapolate) the line to larger and smaller heights.

Does your extended line pass through the origin? _____

The origin represents the limiting case of an imaginary person with height = 0 cm and weight = 0 pounds.

If you imagine arbitrarily small people, do you think that a person
with height = 0 cm should also have weight = 0 pounds? i.e. do you
think that your relationship should pass through the origin?

_____

What does your extrapolated empirical relationship predict
for the weight of a person that is 50 cm high?

_____

What does it predict for a person 250 cm high?

_____

A 50-cm person is equivalent to a new-born baby which typically weighs
about 7.5 pounds. A 250-cm person would be 8'2" tall. What does this
tell you about extrapolating relationships?

Person	Height (cm)	Weight (lbs)	Person	Height (cm)	Weight (lbs)
A			N		
B			O		
C			P		
D			Q		
E			R		
F			S		
G			T		
H			U		
I			V		
J			W		
K			X		
L			Y		
M			Z		

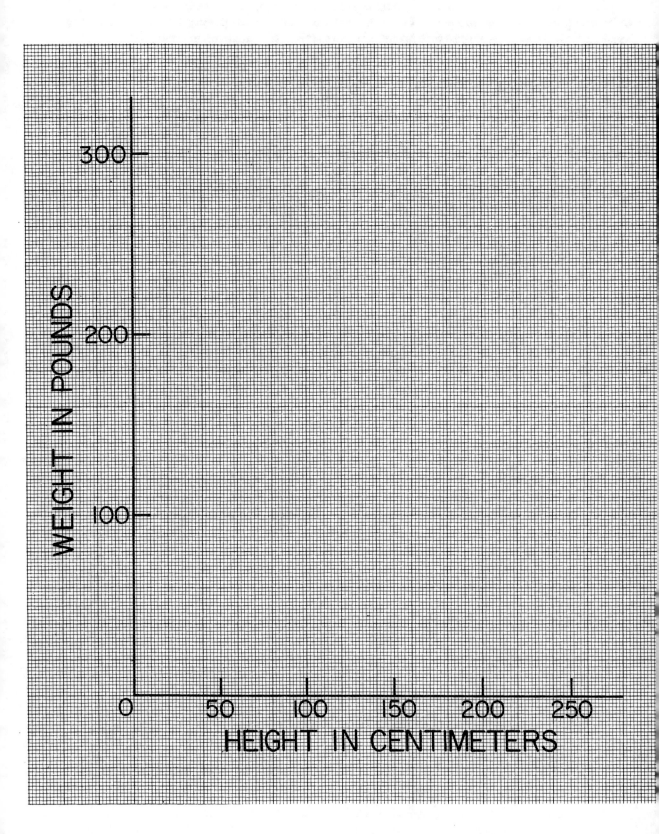

WEIGHT IN POUNDS

300

200

100

0   50   100   150   200   250

HEIGHT IN CENTIMETERS

2

The Celestial Sphere

PURPOSE:

1. To acquire an understanding of how the sky appears from the Earth.

2. The acquire an understanding of the apparent motions of the sun, stars and planets, and the relationship of these bodies to the time of day, time of year and your location on Earth.

EQUIPMENT: Farquhar transparent celestial globe, flexible measuring tape.

REQUIREMENTS: This is a two week lab. The lab will thus count twice as much as the others. Answer all of the questions during the two sessions and turn in the lab at the end of the second week. All work with the globe may be done together with your partner but you should each make some of the measurements. Your written answers must be your own.

INTRODUCTION

The <u>celestial sphere</u> is a conceptual "model" of how the universe appears from Earth. We imagine that all of the objects we see in the sky, like the sun, the moon, the planets, and the stars, are on the surface of a gigantic sphere with Earth at the center. Some of these objects, such as the stars, are fixed to the sphere. Others, such as the sun, moon, and planets move along its surface.

This simple concept of a spherical shell is useful because the distances to the individual objects are so large. The only readily observable effect of the distance is in the brightness of the objects. As long as the distances are much larger (say 100 times) than the size of Earth, there are no other effects of distance that are noticeable from Earth. This condition is satisfied for all celestial objects except our own moon. Thus we may consider all of the objects in the sky to be on a spherical surface - the celestial sphere - at some undetermined but large distance from Earth. Go out one night in a dark place and look at the sky. Doesn't it <u>look</u> like a huge inverted bowl with dots of light on it which we call the stars?

Although we know that some celestial objects are closer than others, and the sun and stars are not really on a sphere, astronomers continue to use the concept of the celestial sphere. The reason is that the celestial sphere is a handy way of finding objects in the sky, especially objects that can only be seen with a telescope. We point the telescope to its proper place on the celestial sphere and <u>voila!</u>, the desired object appears in our sights. Until the development of atomic clocks about 25 years ago, monitoring the positions of the stars on the celestial sphere was the most accurate means of timekeeping. The study of the motions of the planets on the celestial sphere was of fundamental importance in the development of the Newtonian theory of gravity, and the branch of science known as physics. Indeed, small deviations from the expected positions of some planets on the celestial sphere have led to the discovery of other planets. A knowledge of the celestial sphere also permits one to determine one's position on Earth, which is particularly important in navigation.

Getting Acquainted with the Farquhar Globe.

The Farquhar is a "globe-within-a-globe" providing a mechanical model of the celestial sphere. The outer globe represents the celestial sphere and the inner, small globe represents Earth, at the center of the celestial globe, there is a small yellow ball to represent the sun moving on the celestial sphere and a moveable ring around Earth which will be used to represent the horizon. Look through the transparent celestial globe past the Earth-globe, to the far side of the celestial globe to view the sky. The Earth-globe is mounted on a rod connected to a knob at the bottom, the Earth-knob, which can rotate Earth. The Earth-knob should only be turned in a clockwise direction when viewed from outside the globe, as if you were tightening a screw. (This is the direction in which Earth actually rotates, rotating the Earth-globe in the opposite direction can disassemble the globe.) The rod represents the axis around which Earth rotates. Rotating Earth by means of the knob is equivalent to holding the knob and turning the celestial sphere. Thus we can consider the daily rotation of Earth and the apparent rotation of the stars around Earth as equivalent motions. The point at which Earth's axis of rotation connects to the celestial sphere at the bottom is called the south celestial pole. If the rod were connected to the globe from the north, we would locate the north celestial pole.

On the Farquhar globe, the knob near (but not at!) the north celestial pole controls the motion of the sun and is thus called the sun pointer knob. Using the knobs, rotate Earth and also make the sun revolve around the celestial sphere. Notice the sun moves only along a well defined line on the celestial sphere. This line is called the ecliptic. The ecliptic corresponds to the plane defined by Earth's orbit around the sun. We know, of course, that Earth orbits around the sun, but as seen from Earth, the sun appears to orbit around Earth. When considering the celestial sphere, it is more convenient to think of the sun as orbiting Earth as Aristotle thought, but we must always keep in mind that it is really the other way. Notice also that the ecliptic is not perpendicular (that is, at right angles) to Earth's axis of rotation. This is because Earth's equator is inclined by 23.5 degrees to the plan defined by Earth's orbit around the sun. It is this tilt that gives rise to the seasons.

Next, you will find that there is a hole in the star globe through which you can insert your hand. Through this hole you can manipulate the horizon ring, the ring surrounding Earth. When you stand outside at night, you can see only half of the stars, the ones that are "up"; the other half are blocked by Earth, i.e. they are "down". The limit of what you see is called the horizon. On the Farquhar globe, the horizon ring represents the limit of what can be seen. Turn the globe so that the horizon ring is parallel to the floor. The ring now represents the horizon for the topmost point on the globe. All of the stars that are above the horizon ring can be seen from that location. Move the horizon ring to different positions, and note that you can manipulate it for any part of Earth. Finally, locate your position on the globe. Turn the entire sphere on its base so that your city is on top. Now manipulate the horizon ring so that it is parallel to the ground with the compass points facing up. The stars above the horizon are the stars visible at night, during at least part of the year, from your location.

With the horizon set for your location, turn the knob that controls Earth. Remember that rotating the Earth-knob is equivalent to holding the knob and rotating the celestial sphere. You will notice that sometimes the sun is above the horizon and sometimes it is below. Not surprisingly, the time when it is above the horizon is called day, and the time when it is below the horizon is called night. Because of the brightness of the sun, we cannot see the stars whenever the sun is above the horizon. When the sun is aligned with the horizon on the east, we have sunrise, and on the west, sunset, the sun appears to move from east to west across the sky. Using the Earth-knob only, and not the sun-pointer knob, move the sun from east to west from horizon to horizon. Rotating the Earth-knob represents the daily rotation of Earth on its axis.

When the sun-pointer knob is used to move the sun, notice that the sun passes in front of different stars on the celestial sphere. The different positions of the sun correspond to different times of the year. To understand this, we must think for a moment about Earth's revolution around the sun. Consider the diagram below.

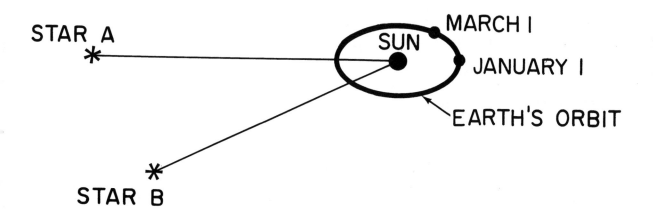

Earth orbits the sun once a year. If we could see stars during the daytime, we would see that on January 1, the sun would appear to be located near Star A. As Earth moves around its orbit, the sun would appear to move along the celestial sphere and be near star B on March 1.

The sun would thus <u>appear</u> to move around the celestial sphere once a year and on each day it would be near a different star. This apparent path of the sun among the stars is known as the ecliptic.

The ecliptic is marked on the globe by dots according to the apparent position of the sun on a given date during the year. Moving the Earth-knob changes the position of the sun relative to the horizon, but does not change it relative to the stars. Since the time of day corresponds to the position of the sun relative to the horizon, turning the Earth-knob changes the time of day. <u>Noon</u> is defined as the time when the sun is on your meridian which is also the time when it is at its highest point above the horizon. The time between two successive noons is divided into 24 hours. An hour corresponds approximately to each black line that connects the north and south celestial poles. These lines are analogous to meridians of longitude on Earth. Turning the sun-pointer knob changes the date, which corresponds to the location of Earth in its orbit around the sun.

The Fixed Stars and the Ecliptic

1. Approximately 1100 stars are represented on the sphere. Yellow dots of seven sizes represent the stars showing relative brightness. What is the name of the brightest star in the constellation Gemini?

    _____

2. The <u>celestial equator</u> is the line around the celestial sphere that corresponds to Earth's equator. Imagine a gigantic slice that cuts through Earth's equator and the celestial sphere. The location of the cut on the celestial sphere is the celestial equator. On what two dates does the sun, moving along the ecliptic, cross the celestial equator?

    _____

    These dates are called the <u>equinoxes</u>. On what dates is the sun farthest from the celestial equator?

    _____

    These dates are called the <u>solstices</u>. The equinoxes and solstices correspond to the traditional dates of the beginning of the seasons.

3. Constellations are <u>apparent</u> groupings of stars on the celestial sphere, associated since the earliest times with mythological characters. Most of these groups are not "real". Constellations boundaries have been determined by international agreement and are shown by blue lines. Through which constellations does the ecliptic pass? With one exception, these are the constellations of the <u>Zodiac</u>. In what constellation is the sun at the vernal (spring) equinox?

    _____

To locate objects on the celestial sphere, coordinates similar to longitude and latitude are used. <u>Right ascension</u> (like longitude) is measured along the celestial equator; circles called <u>hour circles</u> are printed on the globe every 15° (15° = 1 hour). Right ascension (RA) is generally measured in units of hours and minutes. <u>Declination</u> (like latitude) is measured northward and southward from the equator; <u>parallels of declination</u> (Dec) are printed every 10° on the globe. Examples of stars:

	<u>RA</u>	<u>Dec</u>
Betelgeuse	5^h52^m (88°)	+ 7°
Vega	18^h35^m (279°)	+39°

4. Find the RA and Dec for:

	RA	Dec
Summer Solstice	_____	_____
Autumnal Equinox	_____	_____
Winter Solstice	_____	_____
Vernal Equinox	_____	_____
Polaris	_____	_____
Sirius	_____	_____

5. At first glance, the RAs and dates of the solstices are very uncertain because the declination of the ecliptic is roughly constant for about 1 hour of RA which is equivalent to the month of the sun's motion along the ecliptic. How can you estimate the RA of the solstices more precisely than by just looking for the point of largest declination? Hint: Look at the symmetry where the ecliptic crosses various declinations.

The Sky from Your Location Tonight

A few additional definitions.

   <u>Zenith</u>: the point on the celestial sphere that is directly overhead. It corresponds to the point on the celestial sphere that is intersected by a line between the center of Earth and your location on the surface. Every point on the surface of Earth has a different zenith, and the point on the celestial sphere corresponding to the zenith for any specific observer is continually changing as Earth rotates.

   <u>Meridian</u>: the line on the celestial sphere. It is an imaginary line that goes from the north point on the horizon through the north

celestial pole and through our zenith to the south point on the horizon. It, too, is constantly changing with respect to the celestial sphere as Earth turns. The meridian is the place where stars reach their highest point in the sky during the night.

Altitude: the angular distance measured vertically above or below the horizon to a given object. The zenith has an altitude of +90°. The horizon has an altitude of 0°. Altitudes below the horizon are negative. Your instructor will show you how to use a tape measure to determine the altitude of an object on the celestial sphere.

Azimuth: The "compass direction" toward an object measured eastward around the horizon from North. The direction N has azimuth 0° (or 360°), NE has azimuth 45°, SSE has azimuth 157°5, W has azimuth 270°, and so on. The meridian has azimuth 0° for its northern half and 180° for its southern half. Your instructor will show you how to use the "horizon ring" and a tape measure to determine the azimuth of an object on the celestial sphere.

Setting the Globe for a Specific Geographic Location and Time

For the following questions, you will start by setting the globe for noon at your location. Your instructor will tell you the longitude and latitude.

City: _____ Longitude: _____ Latitude: _____

A. With the sphere resting on its base on a table in front of you, position it so that the hand hold is facing you.

B. Rotate the celestial globe and/or the terrestrial globe until your location is "on top". Since no matter where you are on the surface of Earth, you can always think of yourself as "on top"; up is directly overhead (the zenith) and down is toward the center of Earth (the nadir).

C. Reach through the hand-hole and rotate the horizon ring until it is horizontal (parallel to the table) with the labels on the ring facing up. This will eave the gap in the ring (the south edge) nearest you, west on the left and east on the right.

D. At this point, your meridian of longitude should bisect the gap in the horizon ring. Furthermore, the altitude of Polaris above the horizon should be equal to your latitude. For Polaris (but not for any other star) you can use the parallels of declination to "count" up to the altitude.

E. Set the sun so that it is at the desired date.

F. Set the globe for 12:00 noon by holding the Earth-knob fixed and rotating the sky until the sun is to your meridian.

+---+
| Have your instructor check that your globe is set correctly |
| before proceeding further. |
+---+

6. Which constellation is at the zenith at noon?   _____

7. Which named star is closest to the zenith at noon?   _____

8. What is the sun's altitude at noon?   _____

9. Which direction would you face to see the sun at noon? _____

    To set the globe for other times of the day, remember that Earth turns 15° per hour from West to East (15°/hr x 24 hr = 360°). Equivalently, the whole sky moves 15° per hour from East to West. It is easiest to visualize phenomena if you hold the Earth-knob fixed and rotate the celestial sphere to simulate Earth's rotation.

    While counting the circles of RA, rotate Earth eastward (or sky westward) until sunset (the sun at the horizon ring). The number of circles of RA through which you have turned the sky (or Earth) is the number of hours after noon.

10. How many hours after noon does sunset occur (to the nearest 1/4 hour)?   _____

11. What is the sun's altitude now (at sunset)?   _____

12. What direction would you face to watch the sunset today? _____

13. At what longitude is it now noon?   _____

14. Is the constellation you found in question 6 still above the horizon at your location?   (totally, partially, or not at all)

15. Your instructor will choose 2 stars which are just rising.

    a. What are the RA and Dec of each star?
    b. What is the altitude of each star?
    c. What is the azimuth of each star?

Star Name		
Right Ascension		
Declination		
Altitude		
Azimuth		

Now rotate Earth (or the sky) to 3 hours (3 hrs x 15°/hr = 45°) past sunset.

16. What are the altitude and azimuth of the sun?   _____

17. Is the constellation from question 6 still above the horizon? (totally, partially, not at all)

18. What are the altitudes and azimuths of the stars in question 15?

Star Name		
Altitude		
Azimuth		

   Keep the relative orientation of Earth, sky, the sun, and horizon in this same position for the next section of the lab.

The Planets

   Keep the relative orientation of the sun horizon, Earth, and sky fixed in the configuration you set in the previous section!

   The planets are not shown on the celestial sphere because, like the sun, they move around the sky,. Additional mechanical devices like the sun-pointer knob for each planet would be too inconvenient. Your instructor will give your the coordinates for each of the "naked eye" planets. Write these coordinates in the planetary table on this page and mark each position on the globe with a small adhesive dot or piece of tape.

In the table below:

19. Estimate the distance of each planet from the ecliptic in degrees and enter the value in the table.

20. Estimate the altitude of each planet (use negative altitudes for objects below the horizon) and enter the value in the table.

21. Name the constellation in which each planet is found.

Name	RA	Dec	Distance from Ecliptic (N or S)	Altitude	Constellation
Mercury					
Venus					
Mars					
Jupiter					
Saturn					

22. What pattern do you see in the positions of the planets relative to the ecliptic? Describe in one or two sentences. Where on the celestial sphere would you expect the planets to be after 6 months?

Other Times and Places

The appearance of the sky and the rising and setting of objects varies considerably from place to place. Your instructor will give you the locations of one or two other places and a date near one of the solstices. Do the following questions, first for one location and then for the other (if given). Enter all results in the table below.

Set the sun-pointer knob for the given date, set the horizon for the given location, and rotate the Earth-knob (or the celestial sphere) to represent local noon.

23. What are the altitude and azimuth of the sun? Place the value in the table below.

Rotate the Earth-knob (or the celestial sphere) to sunset, again counting hour circles.

24. How long after noon does sunset occur (to the nearest 1/4 hour)? Place the value in the table below.

25. What is the azimuth of (direction to) the sun? Place the value in the table below.

Rotate the Earth-knob (or the celestial sphere) to 3 hours past sunset.

26. What are the altitude and azimuth of the sun? Place the value in the table below.

Date		
Location		
Longitude		
Latitude		
23. Altitude of the Sun		
Azimuth of the Sun		
24. Time to sunset		
25. Azimuth of the Sun		
26. Altitude of the Sun		
Azimuth of the Sun		

27. Name 2 circumpolar constellations (ones that are always above the horizon) at each location.

    A. _____

    B. _____

28. At what latitudes does the sun not rise at all on this date? (Since the sun is highest at noon, adjust the horizon ring to various latitudes holding the sky at local noon.)

    _____

29. Describe in words how the "length of daylight" varies with latitude on Earth near New Year's.

Optional Extra Credit Problems:

30. It is May 20, and it is noon. You are on a sailboat and you use a sextant to measure the altitude of the sun. The sun is to the south and its altitude is 75° What is your latitude?

    _____

31. You are on the same sailboat on July 5, but this time it is late at night. Having just been through a storm, your maps have been washed overboard. You look for the star Vega in the constellation Lyra and notice that it is at your zenith. You have a chronometer (an accurate clock) on board that tells you that it is noon in Greenwich, England.

    In what body of water are you?   _____

    What are your longitude and latitude?   _____


```
* * * * * * * * I m p o r t a n t   N o t e * * * * * * * *

This lab is good preparation for the Independent Observations
     (Lab 15). The sooner you do Lab 15 after this lab is
completed, the easier it will be and the more you will get
     out of it. You should read Lab 15 as soon as you can.
```

5

The Sun and Solar Activity

PURPOSE: To observe and recognize various solar features on several types of photographs, to relate the appearance of features seen against the dark sky to the appearance of the same features in front of the solar disk, and to relate the various features to each other by tracing the influence of sunspots through several layers of the solar atmosphere.

EQUIPMENT: Five sheets of solar photographs provided in class.

REQUIREMENTS: The questions for this lab are to be done individually and should be finished within the lab period.

INTRODUCTION

The sun is a common star, with many millions very similar to it in our Milky Way Galaxy. The sun is unusual only in that it is many thousands of times nearer to us than any other star. Therefore, we can see details and changes on its surface that are invisible on other stars. Some solar phenomena endure just minutes, some days, and some for months.   Though they appear tiny relative to the sun, many of them actually influence our lives, some sporadically, some quite regularly. These phenomena, and how they cause effects on Earth that influence us, are very poorly understood and the subject of much current space research.   It appears that they are all, ultimately, related to electrical currents running in sunspots. The "theme" that unifies this lab is the apparent influence of the sunspots on ever higher layers in the solar atmosphere --- and, indirectly, on us.

Astronomers also study the sun, its temperature, density and chemical composition because we learn something, by analogy, about the many similar stars that populate the Milky Way and even, though we cannot see sun-like stars at those distances, most other galaxies.

This lab is best done after a lecture on the sun or after you read the chapter on the sun in your textbook. In addition, here is some basic information.

The sun is entirely gaseous. It consists mostly of hydrogen and some helium. All the other elements amount only to about 2 per cent of the sun's mass.   The energy used for sunlight is made by nuclear reactions very near the solar center. From there, radiation wanders out slowly, much like light wanders through a dense fog on Earth. Finally it gets to the surface, from where the sunlight can stream out into space. This surface, or "solar atmosphere", is the only part of the sun we can see and photograph.

The solar atmosphere is divided into three (or four) layers because we observe these layers in different ways. See Figure 1 (definitely not to scale). But there are no sharp boundaries. If you were to fly outward through the atmosphere (without evaporating), the gas density surrounding you would simply become ever lower. Near the Earth it is similar to the very best near-vacuum in high-tech laboratories.

Figure 1.

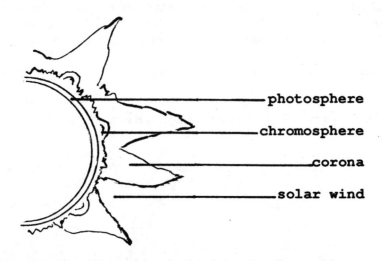

— photosphere

— chromosphere

— corona

— solar wind

The photosphere is that layer of gas from which arrives our ordinary sunlight. The gas emits light of all colors. Its radiation is usually called white light. The white color and the intensity both tell us that the temperature of this layer is about 5700 K. This layer is only about 100 km thick, 1/7000 of the sun's radius. That is why the edge of the sun's disk appears sharp to us.

Sunspots appear as blemishes in the photosphere that are less intensely bright, thus not quite as hot. They carry an electrical current of about a trillion amperes. They act like (gaseous) electromagnets powering all the other phenomena studied in this lab. Short circuits in the currents cause explosions, "flares", whose products endanger astronauts in space, cause northern lights, and may result in widespread electrical power failures.

The chromosphere is the next layer out. Against the dark sky it appears red, hence the name chromo (=color). Its temperature is about 10,000 K. The layer tends to have the shape of flames or loops. The loops, called "prominences" (since they are seen extending beyond the disk of the sun) usually hover over sunspots. Their shapes indicate the spots' magnetism, caused by their electrical currents.

The chromosphere can also be observed in front of the sun's disk when one looks at the sun using only the light of hydrogen or calcium. Photos taken with just this kind of light are called spectroheliograms (spectro- for selecting that specific radiation; helio- for the sun). In spectroheliograms of the sun's disk, gaseous loops appear as dark sinuous filaments. Sunspots evidently influence the chromosphere because the brightest regions in spectroheliograms appear near sunspots.

The _corona_ is the pearly-gray light one sees during a total solar eclipse. It is really a layer of gases at a temperature over a million K. The most easily visible portions of the corona tend to hover over sunspots. Perhaps the electrical currents emanating from sunspots heat the corona to its enormous temperature?

Million-degree gases radiate mostly x-rays. Indeed, x-ray pictures of the sun, taken from satellites, show coronal gases emitting copious x-rays. These gases can be recognized even in front of the sun's disk because the photosphere and chromosphere are not hot enough to emit x-rays themselves. However, regions of strong coronal x-ray emission also hover over sunspots, again indicating the far-reaching influence of sunspots.

The _solar wind_ (the "fourth part of the atmosphere") blows from the sun far into space, past the Earth and all the planets. Humans notice the effect of the solar wind when the wind picks up gases from a comet and drags them out into a long comet tail.

Where the wind starts at the sun, the gases have already been blown away, and so the corona is dark there and there are no x-rays. These regions are conspicuously far from any sunspots.

THE PHOTOGRAPHS

You will be given five sheets of photographs which are reproduced on pages 63 through 67. All were taken on or within a few days of March 7, 1970, a day when a total eclipse of the sun could be observed over most of the US Eastern seaboard. Take a few minutes to identify these pictures:

i)   The picture of the corona (page 66) was taken during the eclipse.

ii)   The x-ray photograph (page 65) was taken shortly after the eclipse. The Moon is still blocking part of the left edge of the picture.

iii)   Two sheets of spectroheliograms (paged 63 and 64) are sketched below.

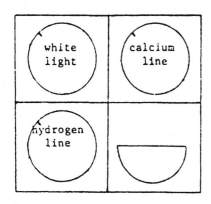

Figure 2.   All March 7 1970

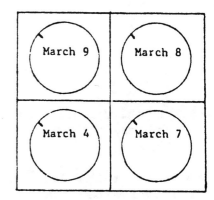

Figure 3.   All hydrogen light

The spectroheliograms are slightly oval because of distortion in the telescope used. The "tick mark" in each picture indicates north as seen from Earth. The upper left picture in Figure 2 is a white-light picture. The other two complete pictures of Figure 2 are spectro-heliograms taken in the light of hydrogen (the label, when reversed, reads Hα) and of calcium (the label, when reversed, reads CaK). [Ignore the half-picture.] The pictures in Figure 3 are all spectroheliograms in the light of hydrogen, taken during several days as shown in the sketch (and in reverse on the labels of the photographs).

iv) The last photograph (page 67) is a magnetogram, recording the magnetization (black or white) of the gases in the photosphere. It was made during several hours of March 7, 1970.

QUESTIONS

1. Using the information provided in the Introduction, identify the layer of the sun's atmosphere shown in each of the following photos:

Figure 2, white light: _____

Figure 2, hydrogen or calcium light: _____

X-ray photograph: _____

Eclipse photograph: _____

2. In which photograph do you expect to find sunspots most easily?

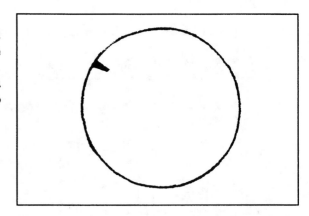

Draw the spots you find as dots at the corresponding places in the circle on the right (Figure 4). Place the dots carefully since you will sketch several more items into the same circle.

Figure 4.

3. The calcium spectroheliogram of March 7 shows some obvious bright regions, called "active regions". Sketch their outlines in Figure 4.

Now look at the hydrogen spectroheliogram of March 7. Can you see the same active regions?

Can you also see sunspots in this picture? Do you recognize the same spots as you sketched in the circle, Figure 4?

What evidence suggests that the brightness (and somewhat higher temperature) of active regions is caused by sunspots?

4. In the same hydrogen spectroheliogram you can see thin dark curvy features, "filaments". Look for the big one shaped somewhat like a horseshoe. That same filament is apparent in each of the other hydrogen spectroheliograms. Observe how the position of that filament changes during the interval March 4, 7, 8, and 9.

Now choose one of the more prominent sunspots of March 7 and find it on at least three of the days March 4, 7, 8, and 9. Is the apparent daily motion of the spot similar to that of the filament?

What phenomenon can explain the observation that both the filament and the spot appear in a different position from day to day?

       We tend to describe the Earth's rotation by referring to its North and South poles or by referring to the Earth's equator, halfway between the poles. Analogously, the sun has an equator. Draw the sun's equator into the circle of Figure 4, using your best estimate for the apparent daily shifts of filament and spot.

Estimate how long it would take the filament and the spot to make one complete rotation with the sun. Show how you arrive at your estimate. Which do you think gives you the better estimate, the filament or the spot? Why?

One rotation takes _____ days because:

_____ gives the better estimate because:

If you did not have a sequence of pictures covering several days but merely a single calcium spectroheliogram, how might you determine the sun's equator? Explain briefly.

How well do the two methods of finding the equator agree?

5. One more kind of activity, also related to sunspots through their magnetism, is the occurrence of large arches of gas that appear beside the sun's disk against the dark sky, the "prominences". In color photographs, they appear as vividly red. Inspect the hydrogen spectroheliograms for evidence that a prominence and a dark filament are actually the same object, merely recognized differently due to the apparently different position as seen from Earth.

The best evidence is in the spectroheliogram of March _____. Describe or sketch the evidence.

6. Now shift your attention to the corona. The goal is to recognize how the coronal structures "attach" to the active regions and sunspots underneath them.

    Compare both the x-ray and the eclipse picture to the calcium spectroheliogram. Which of the two coronal pictures is compared more easily with the picture of the chromosphere? _____

Figure 5, on the right, is a sketch of the x-ray picture. <u>Turn</u> your calcium picture so that its pattern of active regions matches the x-ray picture. Note the direction of the tickmark in the correctly turned calcium picture. As a check for the instructor, <u>draw</u> the tickmark of the calcium picture at its proper location in Figure 5.

Figure 5.

What can you conclude about the relationship between coronal features and active regions and sunspots from a comparison of the two pictures?

Judging from everything you have done so far, what relation would you expect between the x-ray-emitting regions in the corona and sunspots?

7. Now <u>turn</u> the eclipse picture so that it matches the x-ray picture. Match not only streamers to intense x-rays but also that place in the eclipse picture with no corona to that place on the x-ray picture with the least x-ray emission along the rim of the sun. When you have a good match, <u>sketch</u> the largest streamer of the eclipse picture (the one opposite the place where there is no corona at all), first onto Figure 5 and then also onto Figure 4, which already contains your sketch of the sunspots and active regions. (Remember that Figures 4 and 5 have different orientations.)

Judging from everything you have done so far, what relation is there between coronal streamers and sunspots?

Data from the first US space station, Skylab, showed that the place with no corona, furthest from spots, is the source of the solar wind that blows past the Earth. Thus the solar wind's influence on Earth is ultimately also due to sunspots, or rather the lack thereof.

Some coronal streamers may "attach" to active regions behind the sun. Which streamer in the eclipse picture does not seem to match spots and/or active regions in front of the sun on March 7?

8. Review questions:

a. Why did we have to wait for the Moon to cover the sun's disk to take a picture of the coronal streamers but we did not need the Moon for the x-ray picture?

b. What property of the photosphere "explains" the observation that the photosphere does not radiate x-rays like the corona does?

9. Finally, the magnetogram provides evidence that the spatial relations between spots, active regions, filaments, x-rays and coronal streamers have something to do with magnetism (which arises from electrical currents). Black and white on the magnetogram correspond to opposite polarities (as if there were magnetic north and south poles buried under the solar surface).

   Compare the magnetogram with the calcium spectroheliogram. Describe the relative locations of magnetized regions and active regions.

In the northern half of the sun, the black portion of an active region is to the (right, left) of the white portion. _____

In the southern half, the relation is (the same, reversed). _____

   The magnetogram does not show sunspots, but the magnetic pattern of spot pairs is the same as that of active regions. Eleven years later, both patterns have reversed. Still eleven years later, they have reversed again, back to the original pattern. Thus we speak of a 22-year magnetic cycle. Some aspects of Earth's climate such as droughts and temperatures in the high atmosphere show a 22-year cycle. Perhaps they are indirectly caused by sunspots and their magnetism.

   Compare the location of the horse-shoe shaped filament on the calcium spectroheliogram to the magnetic pattern. Describe the evidence that the location of the filament has something to do with the magnetic pattern.

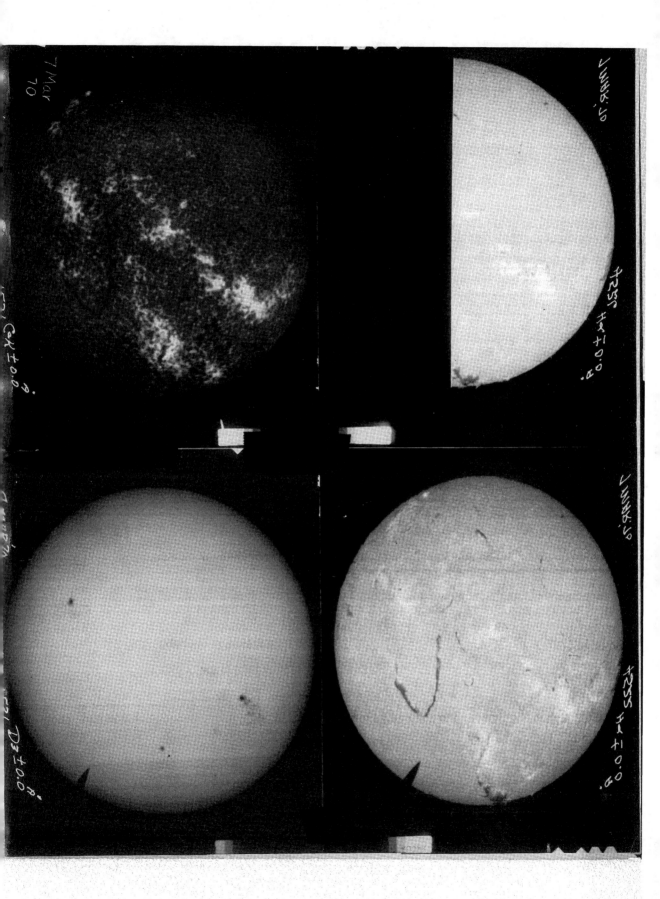

63

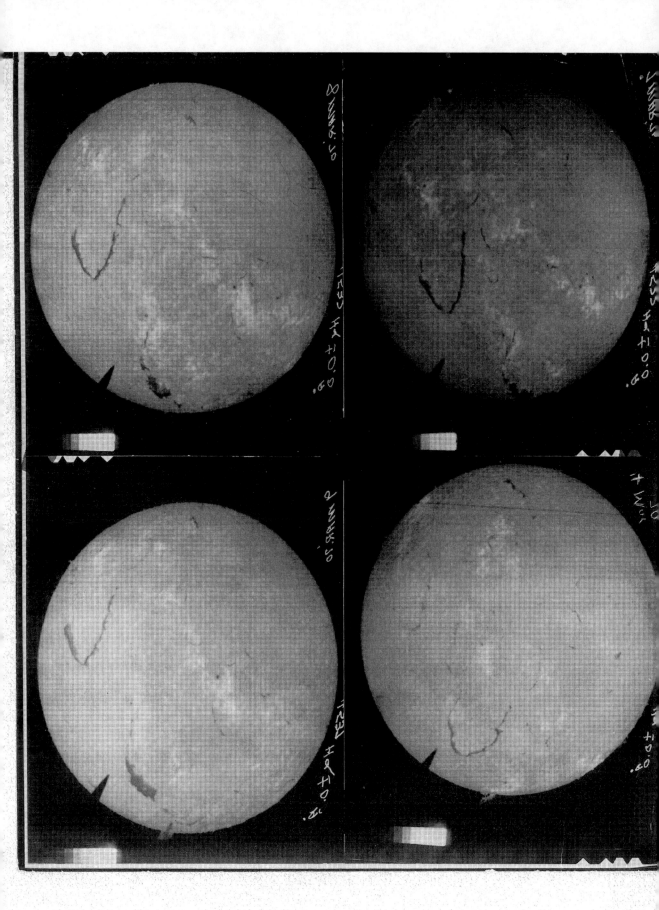

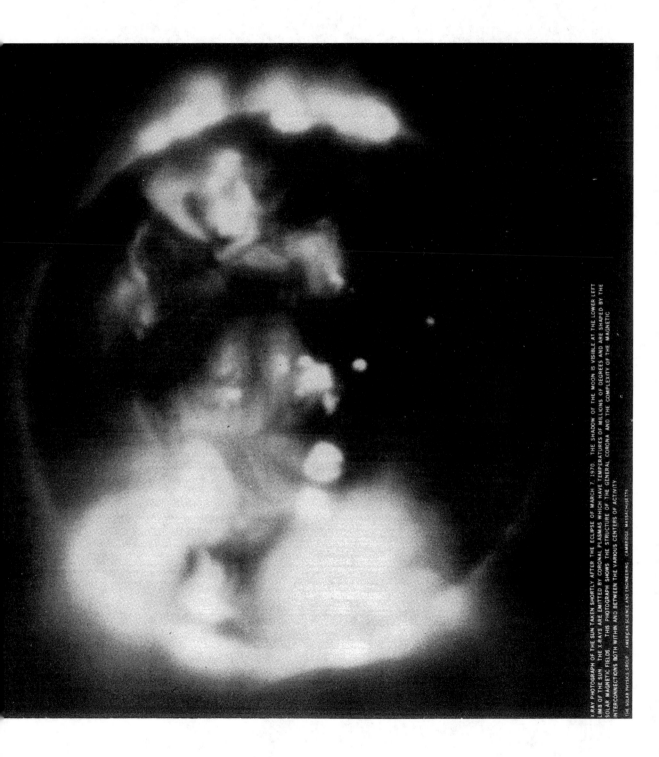

X RAY PHOTOGRAPH OF THE SUN TAKEN SHORTLY AFTER THE ECLIPSE OF MARCH 7, 1970. THE SHADOW OF THE MOON IS VISIBLE AT THE LOWER LEFT LIMB OF THE SUN. THE X RAYS ARE EMITTED BY CORONAL PLASMAS WHICH HAVE TEMPERATURES OF MILLIONS OF DEGREES AND ARE SHAPED BY THE SOLAR MAGNETIC FIELDS. THIS PHOTOGRAPH SHOWS THE STRUCTURE OF THE GENERAL CORONA AND THE COMPLEXITY OF THE MAGNETIC INTERCONNECTIONS BOTH WITHIN AND BETWEEN THE VARIOUS CENTERS OF ACTIVITY.

THE SOLAR PHYSICS GROUP   AMERICAN SCIENCE AND ENGINEERING   CAMBRIDGE, MASSACHUSETTS

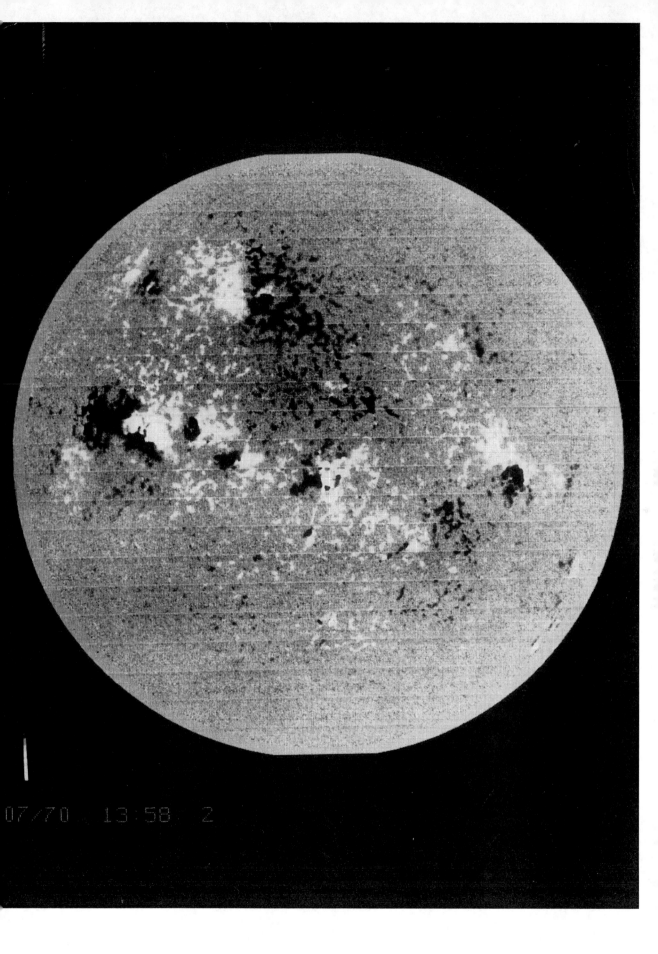

07/70 13:58 2

9

Name _____

Section _____

Lab Partner _____

The Hertzsprung-Russell Diagram

PURPOSE: To construct a Hertzsprung-Russell diagram from a list of the brightest stars and the nearest stars, and to study selection effects in data.

EQUIPMENT: none

REQUIREMENTS: This lab is to be done individually and turned in at the end of the lab.

INTRODUCTION

We wish to consider spectra in a way that is useful for determining both ages and distances of stars. The Hertzsprung-Russell, or H-R, diagram is a widely used tool for astronomers. The H-R diagram is basically a graph of luminosity (expressed as absolute magnitude) versus surface temperature (expressed as spectral type) with the data points being individual stars. As with most graphs, the data could be presented mathematically, but in this case the mathematical presentation would be so complex that astronomers usually confine themselves to the graphical presentation.

In doing this lab you will encounter another of the problems that scientists frequently face, namely, how one should collect data. A scientist must collect data systematically and with a great deal of forethought since basing a conclusion on whatever data happen to be handy can lead to totally misleading results. The reason for this is that the data that just happen to be handy may be subject to strong selection effects. It is the same problem that a pollster faces when trying to predict an election result. If he or she steps out of his or her New York City office and questions only the people who happen to be around, a reliable sample of national opinion will not be obtained.

PLOTTING THE H-R DIAGRAM

Table 1 and Table 2 are lists of the 35 nearest stars, and the 35 brightest stars as seen from Earth. The lists contain the stars names, their distances in parsecs, the spectral types and the absolute magnitudes. In addition, the list of brightest stars also gives their apparent magnitude. After the spectral types you will normally find a Roman numeral called the <u>luminosity class</u> which is a gross classification of the size of the star.

This classification is as follows:

I	– supergiants
II	– bright giants
III	– giants
IV	– subgiants
V	– main sequence stars

Note that in the previous lab we considered only stars with luminosity class V. For plotting the H-R diagram you will need only the spectral class (without the luminosity class) and the absolute magnitude.

Remember that in the system of magnitudes for measuring brightness, negative numbers and small positive numbers represent bright stars while the large positive numbers represent faint stars. On the H-R diagram, the vertical axis is absolute magnitude, with increasing luminosity (decreasing absolute magnitude) going up. The horizontal axis is spectral type with temperature decreasing to the right.

1. The first five stars from Table 1 are plotted as small circles on the H-R diagram on page 109. Label each of these points with the number indicating which star it is, i.e. label the point for the Sun with a small number 1 beside it and the point for Proxima Centauri with a small 2 beside it. Then plot all the other stars from Table 1 as small circles in this same H-R diagram (without numbers).

2. Now plot the stars listed in Table 2 on the same sheet of graph paper; indicate these stars with an x.

3. How many of the nearest stars are also among the brightest stars? _____

   Which stars are these?

4. Which spectral class is most common among the nearest stars? _____
   How do the nearest stars compare to the Sun with regard to luminosity and temperature?

5. Identify the main sequence on the H-R diagram by drawing a line through it. The white dwarf stars are the stars significantly below the main sequence. Identify them by drawing a box around each white dwarf.

6. The hottest stars are typically among the (nearest, brighest) stars. Circle the correct answer.

7. The most luminous stars are typically among the (nearest, brightest) stars. Circle the correct answer.

8. If you made a plot of __apparent__ magnitude versus spectral type for the brightest stars, the plot would look like Figure 1. Does it resemble the H-R diagram? _____ How is it different?

9. Which group, the nearest stars or the brightest stars, has larger apparent magnitudes?

10. For the nearest stars, will the apparent magnitude generally be less than or greater than the absolute magnitude?

SELECTION EFFECTS

Astronomers cannot, in general, change the conditions of an experiment the way a biologist, a chemist or a physicist can. This makes astronomy unique among the sciences. Astronomers must select objects for observation from among those that nature has provided. Because of this, astronomers must always be wary that the objects they have selected have some bias which could lead them to erroneous conclusions. This type of bias is called a selection __effect__.

11. Using Tables 1 and 2, count how many stars there are in each spectral class and place your answer in the table below.

Nearest Stars

Spectral Class	Number	Density
O		
B		
A		
F		
G		
K		
M		

Brightest Stars

Spectral Class	Number
O	
B	
A	
F	
G	
K	
M	

The distribution of stars in these two lists is very different. How are they different?

12. A good way to see the fundamental difference between the two lists is to determine the actual space density of stars for each spectral class, i.e. the number of stars of each spectral type per cubic parsec. The first 33 stars on the list of nearest stars (Table 1) are all the stars within 4 parsecs of the Sun. The volume of space included in this sphere is given by

$$\frac{4}{3}\pi r^3$$

with r = 4 parsecs or about 250 cubic parsecs. Using the set of all stars within 4 parsecs (i.e. eliminating the last two stars from Table 1), calculate the number of stars of each spectral type per cubic parsec and put the result in the table on the previous page. Which type of star has the highest space density? _____

13. Now, let's calculate the space density of the hottest stars. Since the hot stars are intrinsically bright, it is possible to see them over very large distances. Let us assume (actually a good assumption) that in the list of the brightest stars, we have identified all of the B stars within a radius of 50 parsecs. Using Table 2, how many B stars are within 50 parsecs of the Sun? _____ The volume of space contained within 50 parsecs is about 500,000 cubic parsecs. What is the density of B stars within this volume? _____ How does this density compare to the density of any of the spectral classes among the nearest stars?

14. The nearest O star on the list is _____ parsecs away. From this, you would expect that the density of O stars is (greater than, less than) the density of B stars. Circle the correct answer. Using the information from the tables in question 11, one would conclude that hot stars are therefore relatively (common, uncommon) and cool stars are relatively (common, uncommon) in the Milky Way. Most of the stars in the Milky Way are probably stars of what spectral class? _____

15. Clearly the distribution of stars in the two samples is different and this difference is due to the way in which the two sets of stars were selected. Based on the information from this lab, imagine that you made an H-R diagram containing all the 10×10^{11} stars in the Milky Way. Would most of the stars fall on the part of the diagram defined by the nearest stars or on the part defined by the brightest stars? Why?

If you have done this part of the lab correctly, you will have discovered that choosing a certain set of stars as representative of most of the stars in the Milky Way can bias our conclusions about the properties of typical stars. It's somewhat like taking the average age of all of the students in your lab section as representative of the average age of all of the people in the Washington area. In whatever you do, beware of selection effects! In other words, when you pick a sample of objects to study, you want to be very careful how you choose the sample.

TABLE 1 - THE NEAREST STARS[1]

Star	Distance (pc)	Spectral Type	Absolute Magnitude
Sun	5×10^{-6}	G2V	4.8
Proxima Centauri	1.30	M5V	15.4
Alpha Centauri A	1.33	G2V	4.4
Alpha Centauri B	1.33	K0V	5.7
Barnard's star	1.83	M3V	13.2
Wolf 359	2.38	M6V	16.6
Lalande 21185	2.52	M2V	10.5
Luyten 726-8 A	2.58	M6V	15.5
Luyten 726-8 B	2.58	M6V	16.0
Sirius A	2.65	A1V	1.4
Sirius B	2.65	A5	11.2
Ross 154	2.90	M4V	13.1
Ross 248	3.18	M5V	14.8
Epsilon Eridani	3.30	K2V	6.1
Ross 128	3.36	M4V	13.5
61 Cygni A	3.40	K4V	7.6
61 Cygni B	3.40	K5V	8.4
Epsilon Indi	3.44	K3V	7.0
BD +43°44 A	3.44	M1V	10.4
BD +43°44 B	3.44	M4V	13.4
Luyten 789-6	3.44	M6V	14.5
Procyon A	3.51	F5V	2.6
Procyon B	3.51	F8	13.0
BD +59°1915 A	3.55	M3V	11.2
BD +59°1915 B	3.55	M4V	11.9
CD -36°15693	3.58	M1V	9.6
G 51-15	3.60	M7V	17.0
Tau Ceti	3.61	G8V	5.7
BD +5°1668	3.76	M4V	11.9
Luyten 725-32	3.83	M4V	14.1
CD -39°14192	3.85	K5V	8.7
Kapteyn's Star	3.91	M0V	10.9
Krüger 60 A	3.95	M3V	11.9
Krüger 60 B	3.95	M5V?	13.3
BD -12°4253	4.05	M4V	12.1

[1]Adapted from the 1989 Observer's Handbook (Royal Astronomical Society of Canada)

TABLE 2 - THE BRIGHTEST STARS[2]

Star	Distance (pc)	Spectral Type	Absolute Magnitude	Apparent Magnitude
Sun	5×10^{-6}	G2V	4.8	-26.8
Sirius	2.7	A1V	1.4	-1.5
Canopus	30	F0I-II	-3.1	-0.7
Alpha Centauri A	1.3	G2V	4.4	-0.1
Arcturus	11	K2III	-0.3	-0.1
Vega	8.0	A0V	0.5	0.0
Capella A	14	G1II	-0.7	0.1
Rigel A	250	B8I	-6.8	0.1
Procyon A	3.5	F5IV-V	2.6	0.4
Betelgeuse	150	M2I	-5.5	0.4
Achernar	20	B5V	-1.0	0.5
Beta Centauri A	90	B1III	-4.1	0.6
Altair	5.1	A7IV-V	2.2	0.8
Alpha Crucis A	120	B1IV	-4.0	0.9
Aldebaran A	16	K5III	-0.2	0.9
Spica	80	B1V	-3.6	0.9
Antares A	120	M1I	-4.5	0.9
Pollux	12	K0III	0.8	1.2
Fomalhaut A	7.0	A3V	2.0	1.2
Alpha Centauri B	1.3	K0V	5.7	1.3
Deneb	430	A2I	-6.9	1.3
Beta Crucis	150	B0IV	-4.6	1.3
Regulus A	26	B7V	-0.6	1.4
Adhara A	200	B2II	-5.0	1.5
Castor A	14	A1	0.9	1.6
Shaula A	100	B1V	-3.4	1.6
Bellatrix	93	B2III	-3.3	1.6
Gamma Crucis	70	M3II	-2.5	1.6
El Nath	55	B7III	-2.0	1.7
Miaplacidus	26	A0III	-0.4	1.7
Alnilam	470	B0I	-6.7	1.7
Al Na'ir	21	B5V	0.2	1.7
Alioth	25	A0	-0.2	1.8
Dubhe A	32	K0III	-0.7	1.8
Alnitak A	450	09.5I	-6.4	1.8

[2]Adapated from Allen, Astrophysical Quantities

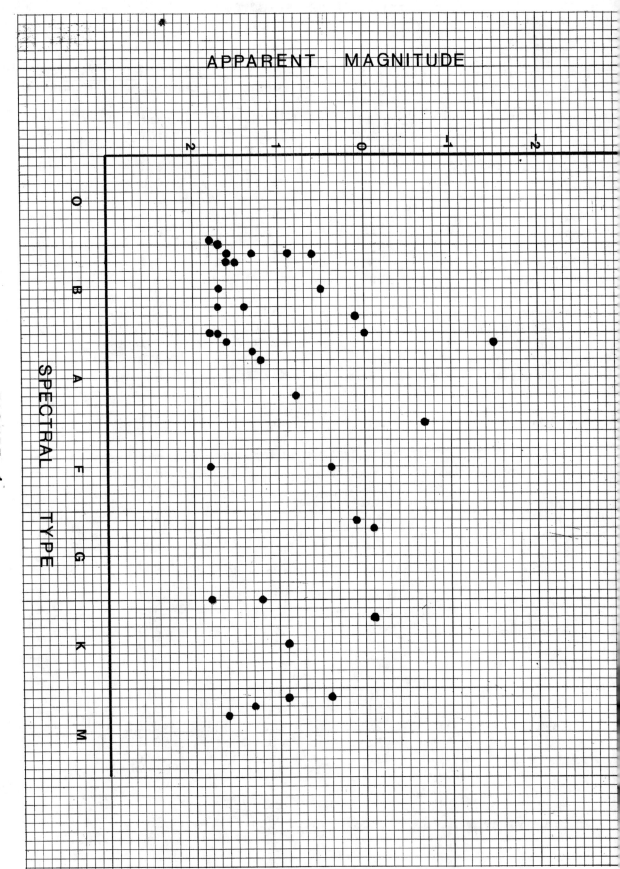

FIGURE 1

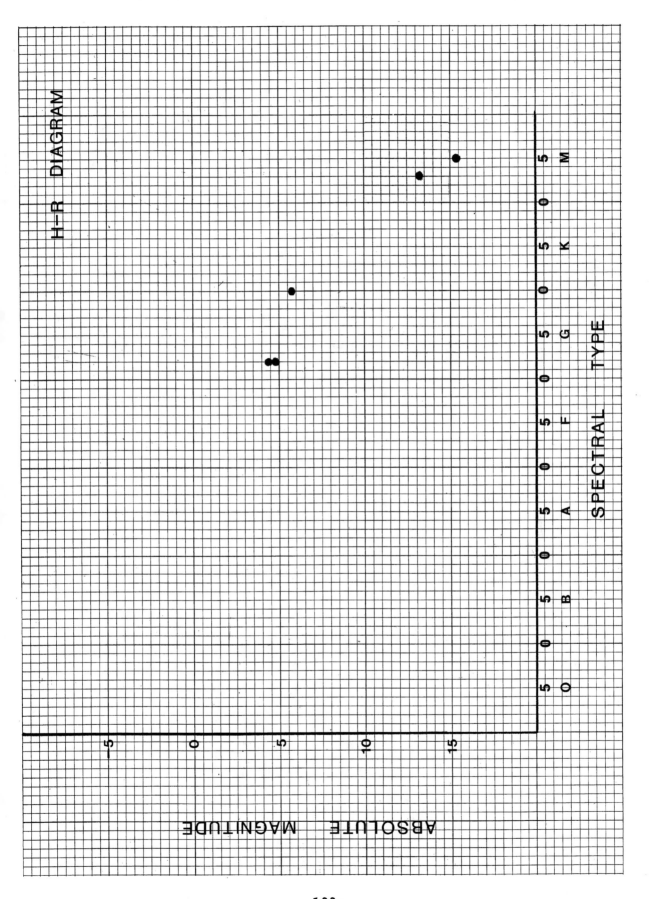

10

Name _____

Section _____

Lab Partner _____

The Crab Nebula

PURPOSE: To measure the expansion rate of the remnant of a supernova explosion known as the Crab Nebula, and to use the expansion rate to determine when the explosion took place.

EQUIPMENT: A ruler for each student in the class.

REQUIREMENTS: This lab is to be done individually. Answer all questions on the worksheet at the end of the lab.

INTRODUCTION

In this lab you will study the most prominent remnant of a supernova in the sky. The Crab Nebula is thought to be the remnant of a supernova that was recorded by Chinese and Japanese astronomers. The nebula emits radio waves, gamma rays, and x-rays as well as visible light. At the center of the nebula is a pulsar which has also been observed to pulse optically -- one of only two pulsars observed in visible light. The nebula has been carefully photographed for most of this century and we will use some of these observations to study the expansion of the nebula and from this to find both its distance and its age.

Look at the photograph of the nebula in Figure 1. It looks like a slightly elongated blob which expanded from a point near the center of the photograph.

DISTANCE OF THE NEBULA

It is convenient to think of the nebula as an idealized spherical shell which is expanding uniformly outward from the supernova itself (the nebula was, of course, thrown off by the supernova when it exploded). Because of the expansion, the front side moves directly toward us and shows a Doppler shift to the blue. The back side of the nebula, on the other hand, moves away from us and shows a Doppler shift to the red. The "sides" of the nebula expand across our line of sight; they move neither toward nor away from us and thus show no Doppler shift. The sides, however, appear to move across the sky so that we see an actual expansion of the nebula if we wait long enough. This motion is called <u>proper motion</u>.

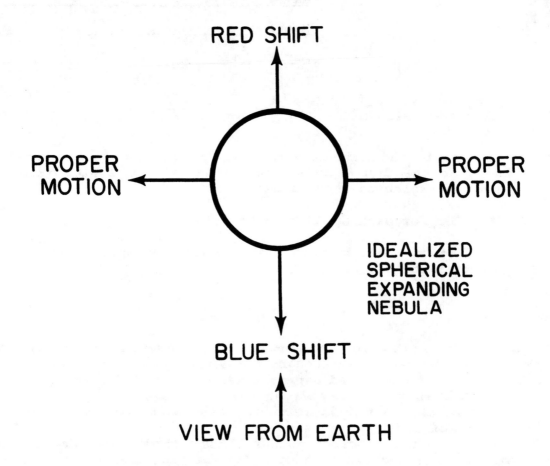

The proper motion of any astronomical object is related both to its speed and distance from the earth. Proper motions in astronomy are usually very small and are measured in seconds of arc per year. For example, if the position of a star or of a filament in the Crab Nebula is found to change its position on a photograph by 1 second of arc in 10 years, its proper motion is 0.1 seconds of arc per year. This is usually written 0.1"/yr. Consider the diagram below.

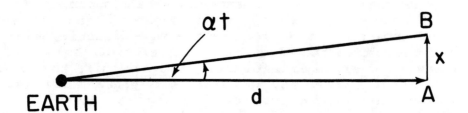

Imagine that an object at A has a velocity v in kilometers per year. In t years, it moves to B, and the distance x it has traveled is

$$x = vt. \qquad (1)$$

From Earth, we see only that it has changed its position relative to some background stars, and we measure a proper motion α. Since α is the rate at which the object changes its position per year, the total change in t years is given by αt. Now, α is proportional to the velocity, v, (the larger v is, the larger α is and inversely proportional to the distance) d, (the larger d is, the smaller α is). Since α is always a very small angle, we may write that

$$\alpha t = \frac{x}{d} = \frac{vt}{d}$$

or

$$\alpha = \frac{v}{d}$$

(2)

if α is expressed in the proper units. $x = vt$ can be considered to be very nearly equal to the arc of a circle centered at the Earth and going through points A and B. There are 1.30×10^6 seconds of arc in a full circle. Therefore, the fraction of a circle represented by AB, if α is measured in seconds of arc, is $\alpha t / 1.30 \times 10^6$. Therefore, in these units we can write

$$\frac{\alpha t}{1.30 \times 10^6} = \frac{vt}{2\pi d}$$

or

(3)

$$\alpha = 2.0 \times 10^5 \, \frac{v}{d}$$

Since we can measure α directly, if we have some way of knowing v, we can determine d, the distance to the object.

We first look at the apparent angular expansion of the nebula using the direct photograph of the nebula with the arrows superimposed on it. To measure the expansion we would, in principle, take a photograph of the nebula, wait twenty years, take another photograph, measure the size of the nebula on each of the photographs, and compare them to see how much the nebula has expanded. Unfortunately, even over twenty years the amount of expansion is quite small and can only be measured accurately with very precise equipment used by an experienced observer. What we have done, therefore, is to take the results of precisely that experiment, calculate how much each part of the nebula should move in 500 years, and show this motion on the photo with an arrow. The original photograph, which is reproduced in Figure 1, was taken in 1938. The tips of the arrows, then, represent the predicted size of the nebula in the year 2438.

1. Using the photograph of the Crab Nebula, measure the expansion of the different parts of the Nebula by measuring the lengths of the arrows. Note that the nebula is not really expanding uniformly so that the arrows are not all the same length. Enter this data, all subsequent data, and the results of calculations on the worksheet.

Since we will be interested only in the average expansion, calculate the average of your data. Because these arrows represent the expansion in 500 years, divided by 500 to get the expansion per year.

The linear distances on the photograph (in millimeters) must be related to the actual angular measurements in the sky (in seconds of arc). This is usually done by taking a photograph of two objects whose angular separation is known. This has been for you, and in the lower left corner of the photograph is a line representing the length of 100 seconds of arc. This is about 27 mm long so we know that each mm of the photograph corresponds to 3.8 seconds of arc in the sky.

2.   Since each mm corresponds to 3.8 seconds of arc, convert your average expansion rate found in Question 1 from mm/yr to arcsec/yr by multiplying by 3.8 arc sec/mm. This is the proper motion, α, described above.

The next step is to determine the linear velocity of expansion, v. This is measured from the Doppler shift of the front and back parts of the nebula. Figure 2 shows a small portion of the spectrum of the nebula with blue on the left and red on the right. To take this photograph, a telescope was pointed so that the nebula produced an image in the slit of the spectrograph. This slit serves the same purpose as the slits on the spectroscopes you used in an earlier lab. Thus the spectrum refers to only a narrow strip across the center of the nebula. The top of the spectrum is due to the light from one edge of the nebula, the bottom of the spectrum is due to the light from the other edge of the nebula, and the center of the spectrum will be the light from the center of the nebula. If the nebula is not uniform in every way, the spectrum will not be uniform either. In fact there can be large differences from one part of the nebula to another and there can be correspondingly large differences in the spectrum from top to bottom.

The nebula emits most of its radiation as distinct emission lines as do most nebulae and the spectral lamps we used a few weeks ago. For this experiment, we have selected just a single emission line from the spectrum, but the entire spectrum would consist of a large number of lines looking very much like the one you are considering. You might expect that a single emission line would just give a single vertical line in the spectrum but this is not the case. First of all, some parts of the nebula are very faint so the spectral line contains both bright and faint patches corresponding to the bright and faint parts of the nebula. Second, one must allow for the Doppler effect. The edges of the nebula are moving sideways across the sky, neither toward nor away from the earth, and thus show no Doppler shift. The center of the nebula, however, is expanding toward us and thus shows a Doppler shift toward the blue. One would expect therefore to find not a straight vertical spectral line but one which has its center bowed out toward the blue by the Doppler shift. Because the nebula is more or less transparent, we can also see the back side of the nebula which is expanding away from us and is therefore Doppler shifted to the red. We thus find that the spectral line has two bows in its middle - one to the red and one to the blue. This is shown schematically below.

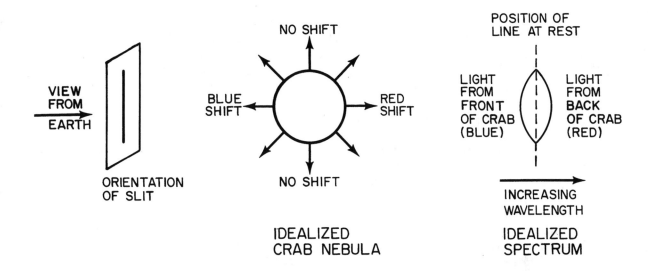

VIEW FROM EARTH

ORIENTATION OF SLIT

NO SHIFT

BLUE SHIFT

RED SHIFT

NO SHIFT

IDEALIZED CRAB NEBULA

POSITION OF LINE AT REST

LIGHT FROM FRONT OF CRAB (BLUE)

LIGHT FROM BACK OF CRAB (RED)

INCREASING WAVELENGTH

IDEALIZED SPECTRUM

Rather than measure the actual Doppler shift of the individual bows of the spectral lines, it is somewhat easier to measure the relative Doppler shift between the two bows of the spectral line. From this you will then calculate the relative velocity between the front and back sides of the nebula which is, of course, twice the expansion velocity.

3. Measure the separation between the bows (see Figure 2) in mm at a few points near the center of both in the photograph and list them in the spaces provided. Remember that the two spectra are for different portions of the nebula so they should give similar but not necessarily identical results. The spectral lines are rather wide, so try to measure from the center of one bow to the center of the other. Take an average of the measurements. The original spectra, from which these photographs were made, were many times smaller so that the astronomer who actually carried out these measurements had to make the measurements using a microscope.

Again, our instrument must be calibrated and this is done with the aid of the comparison spectra which are the four vertical lines above, below, and between the two nebular spectra. These comparison spectra were produced by a spectral lamp like the ones used earlier in the semester. The wavelengths of these lines are accurately known, and by measuring the positions of these lines you can calibrate the spectra of the Crab. This has already been done and we find that each millimeter corresponds to a wavelength difference of 6.76 angstroms.

4. Convert your result from Question 3 to a real Doppler shift (i.e. the actual change in wavelength of the light) by multiplying your result by 6.8Å/mm.

To determine the velocity of the nebula needed in equation (3), use the Doppler formula

$$v = c\frac{\Delta\lambda}{\lambda}$$

where c is the velocity of light, $\Delta\lambda$ is the Doppler shift in Å (found in Question 4) and λ is the unshifted wavelength of the spectral line used (3727 Å $\sim$ 3700 Å). Note that $\Delta\lambda$ and λ have the same units so that their units cancel each other and the velocity will come out in the units we use for the velocity of light. Ultimately we will want velocity in units of km/yr; in these units c = 9.5×10^{12} km/yr.

5. Calculate the relative velocity between front and back of the nebula by using the Doppler formula. Since the expansion velocity is measured from the center of the nebula, divide your results by two.

6. You now have all of the information you need to calculate the distance to the Crab Nebula. Insert the necessary values into equation (3). Remember that the proper motions were calculated in "/yr, therefore the appropriate unit of time is one year. The units of your result will be in kilometers. There are 3.1×10^{13} kilometers per parsec.

What is the distance to the Crab nebula in parsecs? Could the distance to the Crab be measured by parallax? Is it in our own Galaxy? (The Galaxy extends about 15 kpc in the direction of the Crab.) If one parsec = 3.26 light-years, what is the distance to the Crab in light-years?

7. From the measurements of Question 1, the lengths of all of the arrows were not the same indicating that the nebula is not quite expanding uniformly. Using the longest and shortest arrows you measured, how much faster and slower are the fastest and slowest filaments moving with respect to the average? Express your result in terms of a percentage of the average.

AGE OF THE NEBULA

It is now possible to measure the time at which the supernova erupted, a fairly straightforward task using the information you have already measured. In Question 1, you estimated the average expansion rate of the nebula in units of mm on the photograph per year. In the time since the supernova erupted, the material has moved from the supernova itself out to its present position. The star which is thought to be the remnant of the supernova is indicated by the small white spot inside the nebula. The distance the nebula has expanded since the supernova erupted will be given by the distance from the white spot to the base of each arrow.

116

8. Measure the distance from the white spot (representing the star) to the base of each arrow. Enter this data in your table and again calculate the average.

    You now know how far the nebula has expanded since the explosion (Question 8) and how fast it is going (Question 1). We can then divide the distance it expanded by the speed at which it expands to find the time it has taken to expand. This is, of course, just the age of the nebula. We assume, of course, that it has not sped up or slowed down.

9. Calculate the age of the Crab Nebula and subtract this age from the date of the photograph (1938) to determine the year in which the supernova erupted.

10. Oriental astronomers observed a new star, a supernova, at this location in the year 1054 A.D. Does it seem reasonable to you that the errors and uncertainties in the measurements you made in this lab can account for the difference, if any between the year calculated for the supernova eruption and the ancient observation? Would you therefore conclude that the supernova of 1054 is, might have been, or definitely was not the explosion that triggered the expansion of the Crab Nebula?

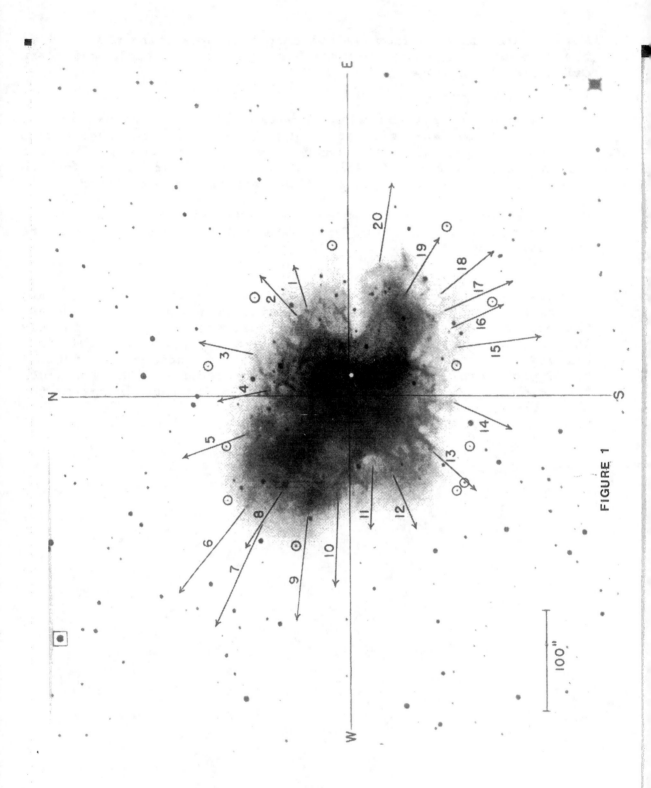

FIGURE 1

100"

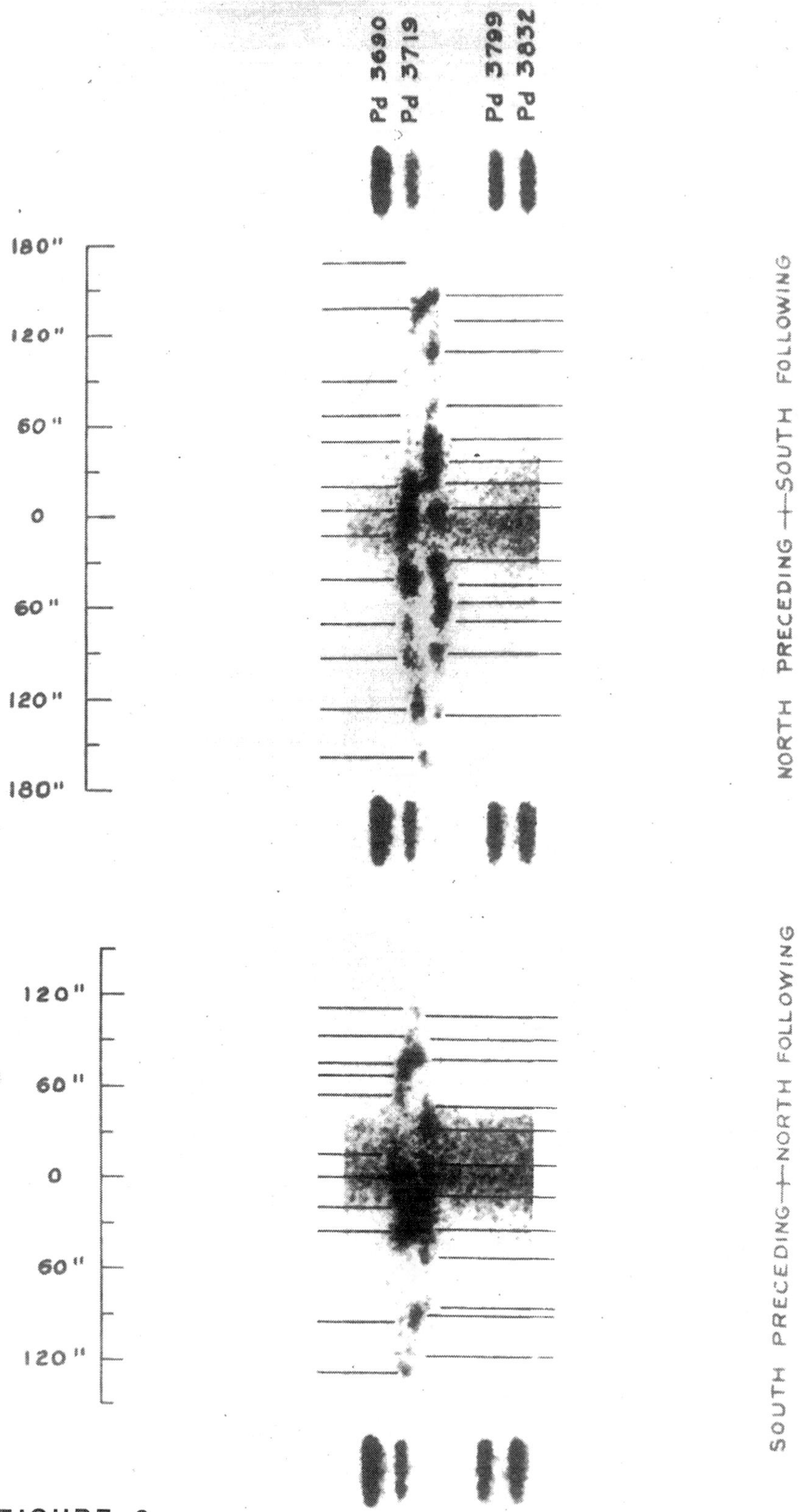

Pd 3690
Pd 3719
Pd 3799
Pd 3832

180"
120"
60"
0
60"
120"
180"

NORTH PRECEDING —SOUTH FOLLOWING

120"
60"
0
60"
120"

SOUTH PRECEDING —NORTH FOLLOWING

FIGURE 2

119

CRAB NEBULA - WORKSHEET

Arrow	Question 1 Length of Arrow (mm)	Question 8 Length to Arrow (mm)
1		
2		
3		
4		
5		
6		
7		
8		
9		
10		
11		
12		
13		
14		
15		
16		
17		
18		
19		
20		

1.   Average arrow length: _____ mm/500 yr

     Average arrow length/500: _____ mm/yr

2.   Average expansion α = _____ seconds of arc/yr.

3. Separations measured between bows:

_____ mm _____ mm _____ mm

_____ mm _____ mm _____ mm

Average: _____ mm

4. Doppler shift between bows: $\Delta\lambda$ = _____ Å.

5. Calculate the velocity from the Doppler formula (Show your work).
   Expansion velocity   v = _____ km/yr.

6. Distance to Crab Nebula: _____ km
   (Show your work)
   _____ pc

   Measurable by parallax? (Yes, No).
   Is Crab in our Galaxy?   (Yes, No).
   Distance to Crab Nebula: _____ light years.

7. Fastest filament as percentage of average _____ .

    Slowest filament as percentage of average _____ .

8. Average length to base of arrows: _____ mm.

9. Age of nebula: _____ years

    Year of explosion: _____

10. Give your answer here to question 10.

12

Name _____

Section _____

Lab Partner _____

Galactic Rotation

PURPOSE: To measure the rotation of the Andromeda Galaxy (M31), determine its mass, and learn how "models" are used in Astronomy.

EQUIPMENT: Computer with galactic rotation program; ruler.

REQUIREMENTS: Data-taking is to be done using the computer together with your lab partner. Other work is to be done individually and handed in at the end of class.

INTRODUCTION

In this lab we take another large step outward in the universe. Everything we have studied up to now has been inside our own Milky Way galaxy. We now look across the vast emptiness of intergalactic space to another nearby galaxy called M31 (number 31 in Messier's catalog) or the Andromeda Galaxy. If we could get outside our own Milky Way galaxy and look back, it would probably look very much like M31. In subsequent labs we will use the distance to M31 to find the distance to other galaxies and even to the edge of the universe.

When we consider the motion of objects, it is often convenient to consider the true velocity (which in our case will be a rotational velocity) as made up of two components because it is often easy to measure one component of the velocity but not the true velocity. The velocity along the line of sight, directly toward or away from the observer, is an easily measured component. For a nearby object, the motion will cause the object to appear to get larger or smaller depending on whether it is approaching or receding. For objects at any distance we can, provided there is enough light from the object, measure the line-of-sight velocity by the Doppler shift, the same effect used by police radar guns. The orthogonal or complementary component of the velocity is the transverse velocity, motion across the line of sight, say left and right or up and down. To measure this velocity we can measure the apparent change of position and then multiply by the distance to the object. In the lab on Jupiter's moons we measured the transverse velocities of the moons (the amplitudes of their motion divided by their periods). For a very distant object, however, the apparent change in position is very small even if the object is actually moving very fast. Think how slowly a jet appears to move when it is 10 miles away in the sky compared to a jet moving down the runway directly in front of you even though the jet on the runway is actually moving much more slowly. For objects outside our own Milky Way Galaxy, we are not yet able to measure transverse motions, with a few rare exceptions, because the motions generally appear so small.

In this lab, and again in lab 14 - Expansion of the Universe, we will measure line-of-sight or radial velocities. In this lab, that line-of-sight velocity is due to rotation of the galaxy so we will use a model of the galaxy's motion to determine the true rotational velocity from the measured line-of-sight velocity. Finally, we will use Kepler's laws to determine the mass of the galaxy from the rotation, much as we obtained the mass of Jupiter from the revolution of the Galilean satellites around Jupiter in lab 4 - the Moons of Jupiter.

ROTATING DISCS

On the opening screen for the computer, select the Galactic! lab with the mouse. The first part of the program demonstrates the motion of a rigid disc like a record or CD. The main part of the display is a representation of a disk with arrows indicating the direction of motion. To the right is a "side view" showing the observer's eye looking at the rotating disc. As you move the cursor around on the disc, displays show the distance from the center of the disc with the mouse, the true rotational velocity at that point, and the line-of-sight velocity which would be measured. Since in this case the disc is being viewed <u>face-on</u> (Inclination = 90°), the line-of-sight velocity is everywhere 0. You will notice that as you move from the center to the edge the velocity increases continuously.

If you click the button on the mouse while the cursor is on ChangeInclination!, you can view the disc at other inclinations. In these cases the line-of-sight velocity also increases steadily from the center of the disc toward either side but not toward the top or bottom. Unless you go all the way to inclination = 0°, or <u>edge-on</u>, the line-of-sight velocity will always be less than the true rotational velocity. The "side-view" box and the arrows on the disc should help you to visualize why this is true.

ROTATION OF THE ANDROMEDA GALAXY

Go to the NextPart! of the lab. The computer screen shows a picture of the Andromeda Galaxy which is a large spiral galaxy, the nearest such galaxy to the Milky Way. A photograph of M31 is on page 140 of the lab manual. Astronomers assume that spiral galaxies are all rotating disks of stars and gas. It is clear that Andromeda does not appear as a circular disk and the most reasonable explanation is that we are seeing it at a fairly large inclination. In fact, its overall shape resembles that of one of the more highly inclined disks shown in the previous part of the lab.

Above the image there is a line with two small filled circles at either end and a small cross to indicate the center. If you move the cursor arrow to either circle or to the center point and click on it, then moving the cursor arrow will move the line on the screen. Position the line so that it runs down the long axis of the galaxy with the cross at the center of the galaxy.

If we were actually doing this experiment, we would take spectra at various points along this line, measure the Doppler shift and thus determine the radial velocity. We would next measure on the photograph the distance from the center to the various points at which the spectra were taken. They would usually be made in some convenient units, e.g. mm. It is next necessary to know how away Andromeda is, so that we can convert the measurements on the photograph (in mm) to actual distances (in kpc). This is obviously a time consuming task and all these calculations have been done for you. Notice that there is a small "bead" which moves along the line as you move the cursor around. This bead marks the point where you are "taking a spectrum" and measuring the radial velocity via the Doppler shift. The readouts on the screen give the velocity of the point at which the bead is located. If you have centered the line on the galaxy, you will be recording the radial distance from the center of the galaxy and the radial velocity of that point. If you have made an error in positioning the line, you will have some error introduced into your readings.

1. Starting at the center of M31, position the cursor at points separated by approximately 3 kiloparsecs and record the distance and line-of-sight velocity, v_{los}. Record the data in Table 1 for values from about 21 kpc in the northeast, through the center, to 21 kpc in the southwest.

Table 1: Worksheet for Rotation of the Andromeda Galaxy

R [kpc]	v_{los} [km/s]	v_{corr} [km/s]	v_{rot} [km/s]
00.0			

Northeast quadrant (upper left)

Center

Southwest quadrant (lower right)

If you now recall the demonstration of the rotating disks (Part 1 of the computer program), you should realize that there is something wrong. For a rotating disk, the center point did not move and had a zero velocity. However, the radial velocity at the center of this Galaxy is far from zero. This is because there are other motions involved besides rotation. Andromeda and our Galaxy may be getting closer together or further away from one another. Our sun is presumably revolving around the center of our Galaxy. The earth is revolving around the sun and rotating on its axis. All of these motions can add to or subtract from the radial velocity due just to the actual rotation of Andromeda. The motions of the sun are reasonably well known and are corrected for in the values you have just measured. However, the overall motion of Andromeda with respect to us is still in the data and has to be determined. The simplest way to do this is to assume that the center of Andromeda is not rotating and that all the motion we see there is systemic i.e. is due to the overall motion of Andromeda.

2. Record the velocity at the center and indicate whether it implies that Andromeda is approaching or receding. You can determine the systemic velocity more precisely by averaging all the velocities as long as the points are symmetrically placed about the "true" center of the galaxy. Determine the average velocity.

Velocity at Center = _____

Approaching or receding? _____

Average velocity = _____

Would you choose the central velocity or the average velocity as the systemic velocity? Explain.

To correct for this systemic motion, you must subtract the systemic velocity from the values you have previously recorded in Table 1. Do this and record the results in the column labeled v_{corr}. (Don't forget that subtracting a negative number is the same as adding a positive one.)

Finally we have to correct for the fact that the galaxy is tilted. To estimate the tilt, recall that our model of the galaxy would be circular if viewed face on (inclination = 90° as in the first part of the lab). A measure of the tilt (actually the sine of the inclination) can be determined by measuring the longest and shortest diameters of the galaxy.

3. Measure the axes with the ruler on the photograph of the galaxy on page 140.

long diameter = a = _____

short diameter = y = _____

y/a = _____

Now refer to the following diagram.

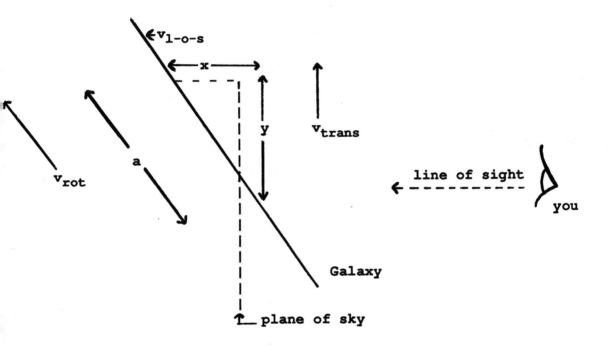

The ratio of the two diameters of the galaxy is just y/a. The measured radial velocity is v_x (i.e. the component of velocity away from you) and the true rotational velocity is v_a.

$$v_a = v_x \times \frac{a}{x} = v_x \times \frac{1}{x/a} = v_x \times \frac{1}{\sqrt{1-(y/a)^2}}$$

4. Thus the tilt-correction factor for the velocities is

$$tilt\text{-}corr = \frac{1}{\sqrt{1-(y/a)^2}} = \underline{\hspace{3cm}}$$

5. Multiply the corrected radial velocities, v_{corr}, in Table 1 by this tilt factor to obtain the true rotational velocities in the galaxy and enter these values in the column headed v_{rot}.

6   Using the true rotational velocity at each radius in Table 1 (column 4), plot a rotation curve for the galaxy on the graph on the opposite page. For the rest of the lab, we will not care which side of the galaxy is approaching and which is receding. Therefore you should ignore the sign of v_{rot} and plot each value as though it were positive but use different symbols for the two halves of the galaxy: X for points in the northeast (upper left) quadrant of the galaxy and O for points in the southwest (lower right) quadrant.

7. Draw one <u>smooth</u> curve which best fits all of the points. Remember that the center should have zero velocity.

8. Is the rotation curve of the galaxy reasonably symmetric? That is, do the points on one side of the center have higher, lower, or about the same velocities as those on the other side of the center?

_____

THE MASS OF THE ANDROMEDA GALAXY

At this stage, you should review what has been accomplished. You started with a two-dimensional picture of Andromeda (i.e. as projected onto the plane of the sky) and information about one component of the velocity (i.e. the motion in the line of sight). A model was assumed, namely that the galaxy was a rotating disk. This enabled you to determine the three dimensional properties of the galaxy. However, the results you have gotten for the rotation curve are quite different from the results in the first part of this lab when you looked at a disk which rotated as a solid body. Keep in mind that in assuming a model you did not assume that Andromeda was rotating as a solid body but merely that it was flat and rotating.

If you examine the portion of the rotation curve (opposite page) near the center of the galaxy, the velocity increases with distance from the center but in the outer parts of the galaxy it does not increase. This means that the outer parts of the galaxy have a longer period of rotation than the inner parts. We also observed this phenomenon in Lab 4 where we saw that Jupiter's outer satellites have longer orbital periods than the inner ones. Let us improve our model for Andromeda by assuming that stars in the galaxy move in circular orbits according to Kepler's third law. That formula, from Lab 4 is

$$M = \frac{R^3}{P^2}$$

where P is the period of a star around the center of the galaxy in years and R is the distance in AU. M, however, is the mass only of that part of the galaxy closer than "R" to the center. (In the case of Jupiter, M was the total mass because nearly all the mass was in Jupiter with

V (km/sec)

R (kpc)

X Northeast quadrant (v > 0)
O Southwest quadrant (v < 0)

almost none in the satellites.) We will use the rotation curve to determine the periods and thus the mass of the galaxy.

9. From the smooth rotation curve drawn through the data on the opposite page, read the values of v_{rot} at 3 kpc intervals and enter them in Table 2.

The distances are already converted from kpc to AU in Table 2 but you need to convert the rotational velocities to periods. The circumference around which a star moves at radius R is just $2\pi R/V_{rot}$. Since we want P in years and we have R in AU, we must convert the velocity to AU/yr from km/sec.

$$P[yr] = \frac{2\pi R[AU] \times 1.495 \times 10\ [km/AU]}{v_{rot}\ [km/s] \times 3.16 \times 10^7\ [s/yr]}$$

This can be simplified to:

$$P[yr] = \frac{R[AU]}{v_{rot}[km/s]} \times 29.7\ [km\text{-}s/AU\text{-}yr]$$

10. Calculate the period for each row in Table 2.

11. Use Kepler's third law, $M = R^3/P^2$, to calculate the mass interior to each distance in Table 3.

Table 2: Rotation Periods

R [kpc]	R [AU]	v_{rot} [km/s]	P [yr]	M [solar masses]
0	0			
3	0.62 x 10^9			
6	1.24 x 10^9			
9	1.86 x 10^9			
12	2.48 x 10^9			
15	3.09 x 10^9			
18	3.71 x 10^9			
21	4.33 x 10^9			

14

The Expansion of the Universe

PURPOSE: To understand the last step of the astronomical distance scale and its implications for the size and age of the universe.

EQUIPMENT: Personal computer with "Expansion" lab exercise.

REQUIREMENTS: Data-taking and other activities on the computer are to be done together with your lab partner. Other parts of the lab are to be done individually.

INTRODUCTION

Near the beginning of the twentieth century, V. M. Slipher measured the <u>radial velocities</u>, i.e. the velocities along the line of sight, of a number of "spiral nebulae". At the time, it was not yet definitely established that the "spiral nebulae" were galaxies like our own but spectral lines characteristic of stars like the sun were recognized. He measured the Doppler shift of these lines and found that the overwhelming majority of "spiral nebulae" had their spectral lines shifted toward the red by large amounts, implying that most of the "spiral nebulae" were moving away from us at very high speeds. By 1929 Edwin Hubble established that the "spiral nebulae" were indeed galaxies, and he was also able to show that the radial velocities were proportional to the distances of the galaxies from the Milky Way.

This was indeed a startling discovery. In a static (unchanging) universe it would be expected that as many galaxies would be approaching us as would be receding from us. Slipher's and Hubble's results showed conclusively that the universe was expanding. One might expect at first glance that the Milky Way is the center of expansion and the galaxies are expanding away from us. However, it is possible to show (though it won't be done in this lab) that the galaxies are all moving away from each other and that the expansion is quite general.

Since the universe is expanding, that is, the distances between the galaxies are increasing, one may consider what the situation was like at an earlier epoch of the universe. Clearly, galaxies must have been closer together. At some earlier time the galaxies must have been so close that the distances between them must have been very small. We can even imagine that there was some time in the past when the distance between them was zero, although the universe surely had a very different character from the universe we see now. You will calculate in this lab how long ago this was true, a time we identify as the origin of the universe.

RELATION BETWEEN RED-SHIFT AND DISTANCE
FOR EXTRAGALACTIC NEBULAE

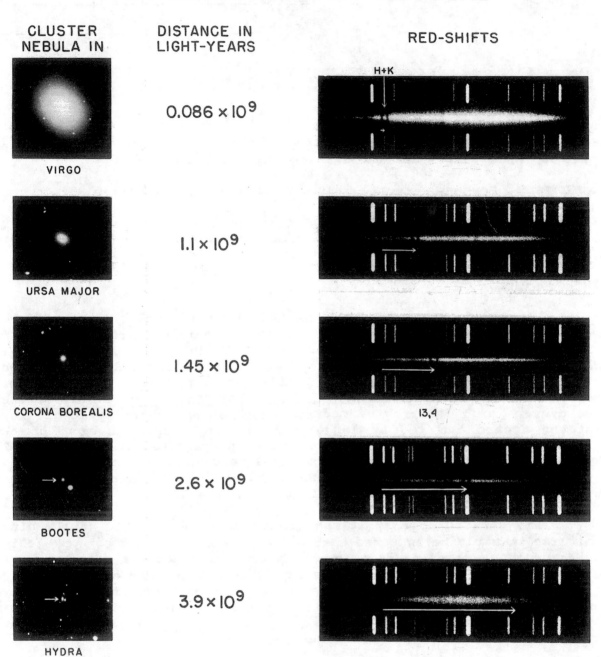

CLUSTER NEBULA IN	DISTANCE IN LIGHT-YEARS	RED-SHIFTS
VIRGO	0.086×10^9	
URSA MAJOR	1.1×10^9	
CORONA BOREALIS	1.45×10^9	
BOOTES	2.6×10^9	
HYDRA	3.9×10^9	

Arrows indicate shift for calcium lines H and K.
One light-year equals about 6 trillion miles,
or 6×10^{12} miles

FIGURE 1

PART 1: CALIBRATING THE SPECTRA

When you start the computer program for this lab, by selecting Expansion! on the start-up menu, you will enter part 1 of the program which shows how the spectra are calibrated, a necessary step before we can determine Doppler shifts from the spectra. The screen displays a simulated spectrum of a galaxy, similar to those in Figure 1. The two strong absorption lines in the galactic spectrum are, for historical reasons, denoted K and H, with K being the one to the left. Now we know that those lines are due to singly ionized calcium and we can look up in a table the wavelengths at which those ions emit and absorb light, 3934 and 3968 Ångstroms (Å) respectively. These lines are the same ones that are prominent in the spectra of cool stars that you studied in Lab 8 – *Stellar Spectra* and they are prominent in these spectra because the galaxies contain many cool stars.

Above and below the spectrum of the galaxy are spectra of a comparison lamp which produces emission lines (analogous to the lamps you studied in Lab 7 – *Spectroscopy*) and which was mounted inside the spectrograph. This lamp vaporized iron in an arc much like the arc that occurs during welding and the spectral lines are used to calibrate the wavelength of the galactic spectrum, i.e., to provide a scale of wavelength <u>versus</u> position. If you position the cursor on any comparison line and click the button on the mouse, the display will highlight that particular line in color and display its wavelength. If you were to drag the vertical crosshair to the spectral line you could read its position, say in mm from the left edge of the scale. Below all the spectra is a scroll bar with a white box. If you position the cursor on the box and click the button on the mouse, you can drag the box back and forth along the scroll bar causing the radial velocity of the galactic spectrum to vary. The display shows the velocity of the galaxy. Note that the spectra of the comparison lines do not move.

An astronomer would use a traveling microscope with an eyepiece having a reticle much like the vertical cross hair in the computer display. This would be used to measure the position of each line in the comparison spectrum just as you use the scale above the spectra to measure the position, d, of each spectral line. The wavelength of each line would then be found in a table. For the spectra in the computer, this calibration has already been done and the relationship between wavelength and position, d, is given by

$$\lambda[Å] \;=\; (13.95\,[Å/mm] \times d[mm]) \;+\; 3743.6[Å]$$

where we have assumed that the smallest divisions of the rule in the display are separated by 1 mm and we have assumed that zero is at the left-hand end of the rule. Note that the units are given in the formula and recall that 1 Ångstrom unit (Å) is 10^{-8} cm.

PART 2: GALACTIC RADIAL VELOCITIES

Figure 1 shows 5 elliptical galaxies, each in a large cluster of galaxies designated by the constellation in which it is located. The top galaxy is in the Virgo cluster, the same cluster that you studied in the previous lab. The other clusters are much further away. Beside each galaxy is a photograph of its spectrum. If you click the mouse on NextPart!, you will get to Part 2 of the program which displays each of these spectra, one at a time. Choose a galaxy by clicking the mouse on SelectGalaxy and then clicking again on the name of the galaxy that you want to choose. Begin with the galaxy in Virgo.

1. Use the mouse to drag the vertical crosshair to the positions of the K and H lines of calcium. Measure their positions (estimating to the nearest 0.1 mm) on the ruler above the spectrum and record the positions in the worksheet at the end of this section. Do the same for each of the other galaxies.

   According to the calibration derived in the previous section,

   $$\lambda[\text{Å}] = (13.95 [\text{Å/mm}] \times d[\text{mm}] + 3743.6[\text{Å}].$$

Therefore we can write

$$\frac{\Delta\lambda}{\lambda_0} = \frac{\lambda - \lambda_0}{\lambda_0} = \frac{(13.95 \times d) + 3743.6 - \lambda_0}{\lambda_0}$$

Where λ, λ_0, and $\Delta\lambda$ are all in Angstroms (Å). Since the wavelengths at which calcium ions emit and absorb can be looked up in a table, we know that $\lambda_0 = 3934\text{Å}$ and 3968Å for the K and H lines respectively. We can therefore rewrite our calibration formula in terms of the Doppler shift:

$$\frac{\Delta\lambda}{\lambda_0} = (0.003547 \times d) - 0.0484, \qquad \textit{for K}$$

$$\frac{\Delta\lambda}{\lambda_0} = (0.003516 \times d) - 0.0565, \qquad \textit{for H.}$$

2. Use the calibration formulae to calculate the Doppler shift for the K and H lines in each galaxy and record the result in the worksheet at the end of this section.

   Now we can use the Doppler formula,

   $$\frac{\Delta\lambda}{\lambda_0} = \frac{v}{c},$$

where c is the velocity of light (3.00×10^5 km/s), to calculate the line-of-sight or radial velocity for each galaxy. But first, since both spectral lines should have the same shift for a given galaxy, we can improve the precision of our result by averaging the Doppler shifts for the two lines. We can then use the average Doppler shifts, $\Delta\lambda/\lambda_0$, to calculate the radial velocity.

3. Calculate the average Doppler shift, $\Delta\lambda/\lambda_0$, for each galaxy and enter it in the worksheet.

4. Calculate the radial (line-of-sight) velocity for each galaxy and enter it in the worksheet.

5. What will be the units of the velocity that you calculate with the Doppler formula? Explain.

Note that the velocities of the successive galaxies are getting systematically larger. If you have been paying attention to the images of the galaxies shown in the upper left corner of the screen and in Figure 1, you will have noticed that the galaxies appear to be successively smaller. This is because they are successively further away. In the next part we will determine the distance.

Worksheet for Radial Velocities

Cluster	Position of Line [mm]		Doppler Shift $\Delta\lambda/\lambda_0$			Radial Velocity [km/s]
	K	H	K	H	Average	
Virgo						
Ursa Major						
Corona Borealis						
Bootes						
Hydra						

PART 3: DISTANCES OF THE GALAXIES

In Lab 13 you determined the distances to the Virgo and Hercules clusters by choosing the four largest spirals in each and comparing their sizes with the size of the Andromeda galaxy, M31. You assumed that intrinsically all the galaxies were similar so that the variation in size was due only to distance. In the next part of the lab (click the mouse on the NextPart! icon to get to Part 3 if you have not already done so) we will use the same technique to determine the distances to the more remote clusters by comparing the size of each galaxy with that of the galaxy in the Virgo cluster. Examine the galaxies in Figure 1. Although you may not be able to tell the types of the galaxies in Bootes and Hydra, all five galaxies are the same type. You should be able to recognize those in the nearer clusters.

6. Are they spirals, barred spirals, or ellipticals?   _____

7. These galaxies should not have the same intrinsic size as the galaxies you used in the previous lab in the Virgo cluster. Does this matter, i.e., will this difference affect your distance determinations? Explain.

In this part of the lab you will measure the sizes more precisely than you did last week. You can again use the mouse to select each of the five galaxies. Begin again with Virgo. Under the galaxy you see a somewhat ragged curve, relatively flat near the center and dropping down to zero at each end. This curve represents the brightness, in arbitrary, relative units of intensity, measured at each point along the major axis of the galaxy. To be consistent in our measurement of major axes, we will always measure between the two points at which the brightness falls to 10% of its value averaged over the center. Beginning with Virgo, center the horizontal crosshair on the "typical" or average value of intensity in the central region of the galaxy and read the value on the vertical ruler to the right of the curve. Record this in the worksheet on the next page. Now calculate 10% of this value and move the horizontal crosshair to that intensity. The two places at which the crosshair intersects the brightness curve are the points we will use to define the ends of the major axis of the galaxy. Move the vertical crosshair to each of these positions, measure the positions on the horizontal ruler below the curve, and record them on the worksheet. The length of the major axis is just the difference between these two values, which sould be entered in the appropriate column on the worksheet.

Follow the same procedure for each of the other galaxies but note that for each of the galaxies except Virgo there is a note on the screen saying "8 x scale". These images have been magnified a factor of 8 in order to make the measurements easier. We will have to correct for this when we calculate the distances.

8. Record all the data measured above in the worksheet on the next page.

In order to have a common starting point, we will assume that Virgo is at a distance of 2.20×10^7 parsecs = 7.25×10^7 light years, a value taken from a recent reference. Compare this with the value you found in the last lab (on the paper returned to you today).

9. Distance to Virgo from Lab 13 _____

Does your measured value agree with the value given above to within a factor of 2? _____ within a factor of 10? _____

To get the distances to the other galaxies, use the same formula that you used last week, but with an extra factor 8 to account for the extra magnification of the images relative to the image of Virgo:

$$d_2 = d_1 \times \frac{8 \, a_1}{a_2}.$$

(Naturally this formula will not work for Virgo itself because the image of Virgo was not magnified by 8.) Inserting the value of $d_1 = 7.25 \times 10^7$ ly for Virgo, we have

$$d_2 = 5.8 \times 10^8 \ [ly] \times (\frac{a_1}{a_2})$$

where a_1 is your measured value for the axis of the galaxy in Virgo and a_2 is your measured value for the axis of any other galaxy. Note that the units of a cancel out as long as both galaxies are measured in the same units.

10. Enter the distances of the other galaxies in the worksheet below.

The distances that you have just derived for these galaxies should be roughly ten times larger than the values derived by Hubble. That is due to the fact that in Hubble's day there were a variety of errors present in the previous steps of the distance scale (the method used to determine the distance to Virgo). Astronomers still do not all agree on the distances to these galaxies because different methods of determining the distances yield answers even today that differ by up to a factor 2.

Worksheet for Distances

Cluster	Brightness B (intensity units)	0.1 x B (intensity units)	Ends of Axis [mm]		Axis Length a [mm]	Distance d [ly]
Virgo						
Ursa Major						
Corona Borealis						
Bootes						
Hydra						

PART 4: EXPANSION OF THE UNIVERSE

You will now examine the results from Parts 2 and 3 to see the expansion of the universe. If you click the mouse on NextPart! to get to Part 4, you will be presented with a table in which you should enter the values for the radial velocity and distance of each galaxy. Put the cursor on the table and you will be able to type a number into the box for the velocity of Virgo (from your worksheet on page 153). Make sure your units are the same as those asked for in the table on the screen. You can press the "Enter" key to move to the next box in the table or you can click the mouse on any box in the table to enter the remaining numbers from the worksheets in Parts 2 and 3 (i.e. page 153 and 155). Powers of ten are entered by typing 'e' followed by the number, e.g. 1.52×10^{10} is entered as 1.52e10. As you enter the numbers, the points will be plotted automatically on the graph on the lower part of the screen. Numbers can be corrected by clicking the mouse with the cursor in the appropriate box, using the backspace key to delete characters, and then retyping.

Once all the numbers in the table are correct, your data should approximate a straight line on the graph. To draw a line through the graph, use the mouse to drag the white ball upwards from the lower right corner of the graph until the line passes through the points appropriately. The lower end of the line has been fixed since we know that our own radial velocity away from ourselves is zero, i.e. the line must pass through the origin. Once you have centered the line, look to the right of the graph and record the value displayed for

11. Sum of deviations = _____ .

The "best" straight line fitting through the data can be found by minimizing this value. Readjust your line until this value has a minimum and record it:

12. Sum of deviations = _____ .

If your data are nearly along the line, your line should not have moved very much from your "eyeball" fit.

This graph is the empirical relationship that Hubble found and is known as Hubble's law. Hubble's law is now used to determine distances to faraway objects for which a Doppler shift (usually called a redshift because the shifts of distant galaxies are always in that direction) can be measured and for which we do not have other measures of distance. All distances to quasars, for example, are measured in this way: we determine the redshift of the quasar and use a relationship like that shown on your screen to determine the distance. This relationship implies that quasars are extremely luminous objects, many times brighter than ordinary galaxies, a fact that is generally accepted now but which was very controversial when quasars were first discovered. Plot the graph from the screen on the blank graph on page 158 including the data for all five galaxies and the straight line fitted to the data.

The slope of the line in your graph is known as Hubble's constant, H, which describes the rate of expansion of the universe. Since the line goes through the origin, you can most easily determine H from the formula

$$H = \frac{v_{rad}}{d}$$

where v_{rad} and d are taken from any convenient point on the line near its uppermost end.

13. What is the value of H? (remember the units!) _____

Over the last 50 years, the value of Hubble's constant has been revised many times and astronomers still argue about a factor of two uncertainty due to the different methods of determining the distances to the galaxies.

14. If all the galaxies in this lab had been half as far away from us, what would have been the value of H? _____

15. The largest velocity that we can ever measure for an object relative to ourselves is the velocity of light, $c = 3.00 \times 10^5$ km/s. Using Hubble's law, what would be the distance of such an object?

_____

Such an object would be at the edge of the universe and astronomers often refer to this distance as the size or radius of the universe.

We can also use Hubble's law to determine the age of the universe. Consider the object in the previous question, moving at the speed of light and at the edge of the universe. If we assume that it has always been moving at that speed, we can easily calculate how long it took to get there from here. Since in general we can write

distance = speed × time,

we can solve for t

16. $$t = \frac{d}{v_{rad}} = \frac{(\frac{c}{H})}{c} = \frac{1}{H} = \underline{\hspace{3cm}}$$

17. Unfortunately, the units seem a bit peculiar. To convert the units of time into something more useful, we need to multiply by the number of km in a light year (9.48×10^{12} km/ly) yielding the age of the universe in seconds. What is this age? _____

18. We can then divide by the number of seconds in a year (3.16×10^7 s/yr). What is the age of the universe in years?

_____

19. Now let's return to Hubble's original determination of the Hubble law. Recall that his original distance estimates were a factor of 10 too small. Assume that all the clusters of galaxies were 10 times closer than they are, but that their radial velocities are unchanged. Recalculate the age of the universe under this assumption.

H _____

Age of Universe in sec _____

Age of Universe in yr _____

How do your answers compare?

157

20. We know from age estimates of globular clusters that they are all about 10-15 billion years old. How does this estimate of the age of the globular clusters compare with the two different ages you calculated for the universe in questions 18 and 19? Are either or both age estimates consistent with the ages of the globular clusters? Explain.

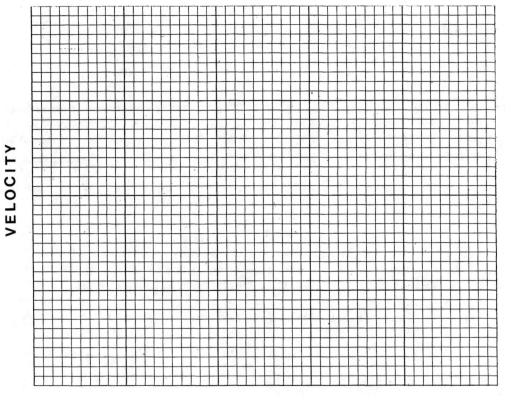

DISTANCE
(in units of 10^9 light years)

15

The Night Sky - Independent Observations

PURPOSE: To introduce you to the night sky and to develop an understanding of angular measurements and rising and setting phenomena.

EQUIPMENT: A star chart, a compass, a flashlight, a wooden pentant. A lab instructor will sign these out to you; you must return them at the end of the night. You will also need to bring a watch.

REQUIREMENTS: This lab is normally done in groups of two, but can be done individually if you prefer. For a period of several weeks after the celestial sphere lab, an instructor will be available every weekday evening at a time specified in the syllabus to sign out the lab equipment and to assist you in getting started. <u>Don't wait until the last week.</u> If the weather is bad that week, you will be out of luck. We frequently have entire weeks where observations are not possible. Even when working with a partner, <u>each</u> partner should make separate observations. The lab must be handed in at the end of the evening session during which you make the measurements.

SUGGESTIONS

A. Dress warmly. Standing around outside, you can get quite cold even if the temperature isn't very low.

B. Don't go out alone. Take a friend even if he or she is not in the class.

C. Pick a clear night. Even thin high clouds will prevent you from seeing all but the brightest stars.

D. Find a location that is a flat open area way from bright lights. There are many such locations on campus. Be sure that you can see the horizon from your location.

E. Read the entire lab before going out to make your observations.

THE EQUIPMENT

A. The <u>Compass</u>. We are providing you with a compass so that you can easily find north at night. Once you learn how to find Polaris (the north star) you won't need the compass again on a clear night.

B. The <u>Star Chart</u>. The star chart will help you find the brightest stars in the sky. To use it, move the wheel so that the date coincides with the time you are observing. In the opening are all of the bright stars you can see above you.

C. The <u>Pentant</u>. The pentant is an angle measuring device that looks like a cross-bow. The curved part is a fifth of a circle (72°). The curved part is a bent yardstick; the important part is the inch scale. To measure the angle between two stars, hold the straight part so that the soft plastic end is against your cheek. Place the beginning of the scale (zero inches) on the first star, and align the pentant so the second star appears just above the inch scale. Each inch = 2°. So, if the second star is seen at 15.5" from the first, the two stars are 31° apart.

ORIENTING YOURSELF

It is first necessary to orient yourself when you go outside. Do this by looking at the compass and facing north. East is to the right, west to the left. Take the star chart (which has been previously set up indoors), hold it parallel to the ground with the North point in the proper direction. You will notice that the east and west on the star chart are reversed. That is because it is meant to be used looking up at it, not looking down. Many people, however, find it easier to look down at the star chart and to transpose east and west.

The North Star is located in the center of the star chart, where the hole is. To find the north star, <u>Polaris</u>, look at the star chart and notice the configuration of stars that make up the <u>Little Dipper</u>. Find the same configuration in the sky and identify the North star. North on the compass will be a few degrees to the west of true north. To check whether you have found Polaris, use your pentant. Polaris will be approximately 39° above the horizon at all times.

In the spring semester, or if you are up very late in the fall, it is easiest to find Polaris by first finding the <u>Big Dipper</u> (refer to your star chart for the configuration of these stars). The two bright stars at the end of the bowl of the dipper (marked pointers on the star chart) point almost directly at Polaris. The two stars are Merak and Dubhe; Polaris is 28° from Dubhe, the closer of the two pointer stars.

OTHER THINGS TO IDENTIFY

In the <u>fall semester</u>, the Big Dipper is very close to the northern horizon and it is often difficult to identify it unless you have an unobstructed view of the northern horizon. The three easiest stars to identify in the fall are the stars known as the <u>summer triangle</u>. These are the three brightest stars in the fall sky and from September to November they are almost directly overhead shortly after sunset. If you go out shortly after it gets dark, look for the three brightest stars overhead; they make a fairly large triangle in the sky. These are the stars <u>Vega</u> (the brightest one), <u>Deneb</u> and <u>Altair</u>. You can see these stars even if the sky is fairly bright from the reflection of city lights.

Vega is the westernmost star, Altair the southernmost. Deneb is the star that is at the top of the <u>Northern Cross</u> (the constellation <u>Cygnus</u>). The cross points to the northeast to another group of stars that is easy to recognize: the constellation <u>Cassiopeia</u>. This constellation has five bright stars in the shape of a bent "W". It is about 30° away from Cygnus. Use your star chart to identify other constellations.

In the <u>spring semester</u>, the easiest constellation to recognize is <u>Orion</u>. Three almost equally bright stars almost directly south 30° or 40° above the horizon, mark the belt. The other stars are easy to find using the star chart, and can be seen even when the sky is bright.

The brightest star in the northern sky, the star <u>Sirius</u>, is to the south and east of Orion. It is in the constellation <u>Canis Major</u>. There are at least six bright stars you should be able to recognize in that constellation. Shortly after it gets dark, passing nearly overhead is a fairly small cluster of stars known as the <u>Pleiades</u>. You should be able to see six stars if your eyes are good and if the sky is clear. You will be measuring the distance to the Pleiades in the lab on stellar spectra. Spring and summer is the best time to view the <u>Big Dipper</u> which is east of Polaris in the spring.

Be sure you can identify all of the objects listed above in the relevant semester you are taking the lab. You will need to make measurements of some of them.

Do this part only if you are taking the course in the fall.

Date: _____ Location: _____

1. Find the bright star near the <u>eastern</u> horizon. Try to identify it on the star chart. What constellation is it in?_____

   Be sure that you can find this star again when you have finished the other parts of the lab.

   How many degrees is it above the horizon? _____ What time is it now? _____

2. What is the angle between Vega and Deneb? _____

3. What is the angle between Deneb and Altair? _____

4. What is the angle between Vega and Altair? _____

5. What is the angle between Polaris and Vega? _____

6. What is the angle between Deneb and the nearest bright star in Cassiopeia? _____

7. Find the stars in a constellation which appears on the star chart but is not mentioned in the introduction. Sketch the positions of all of the stars you can identify on the next page. Measure the angles between the three brightest stars in the constellation and show the angles on your sketch. Measure the angle between the brightest star and the horizon and indicate the angle and the time you made your measurement below the sketch. Be sure to indicate which star you used to make your horizon measurement.

8. Find the star you measured in question 1. How far above the horizon is it now? _____ What time is it now? _____

Finish the rest of the lab inside.

9. How many degrees has the star you identified in question 1 changed in altitude moved since you first observed it? _____
   In how many minutes? _____

10. The star not only moved higher in the sky, but also moved from east to west. You measured only the vertical component of the motion. To find the approximate total number of degrees the star moved, multiply your answer from question 9 by 1.4. (This number is valid only for the latitude of College Park.) Result _____

11. The stars move 360° around the sky in 24 hours (1440 minutes). Over what fraction of a day did you observe? _____

Using the result from question 10, what fraction of a circle did the star traverse? How well do your answers agree?

If you could measure the angles very accurately (and also measure the east-west change in angle accurately) your results should be very close. A good result will differ by no more than 50%. Excellent results will differ by no more than 25%. Accurate observations like these were, until very recently, the most reliable method of timekeeping.

Do this part only if you are taking the course in the spring.

Date: _____ Location: _____

1. Find the bright star near the <u>eastern</u> horizon. Try to identify it on the star chart. What constellation is it in?_____

    Be sure that you can find this star again when you have finished the other parts of the lab.

    How many degrees is it above the horizon? _____ What time is it now? _____

2. The two brightest stars in Orion are Betelgeuse and Rigel. Find them on your star chart and then find them in the sky. What is the angle between them? _____

3. What is the angle that Sirius makes with the horizon? _____

4. What is the angle between Sirius and Betelgeuse? _____

5. What is the angle between the two pointer stars in the Big Dipper?

    _____

6. How many degrees are the Pleiades from Polaris? _____

7. Find the stars in a constellation which appears on the star chart but is not mentioned in the introduction. Sketch the positions of all of the stars you can identify on the next page. Measure the angles between the three brightest stars in the constellation and show the angles on your sketch. Measure the angle between the brightest star and the horizon and indicate the angle and the time you made your measurement below the sketch. Be sure to indicate which star you used to make your horizon measurement.

8. Find the star you measured in question 1. How far above the horizon is it now? _____ What time is it now? _____
    ____

Finish the rest of the lab inside.

9. How many degrees has the star you identified in question 1 changed in altitude moved since you first observed it? _____
    In how many minutes? _____

10. The star not only moved higher in the sky, but also moved from east to west. You measured only the vertical component of the motion.

To find the approximate total number of degrees the star moved, multiply your answer from question 9 by 1.4. (This number is valid only for the latitude of College Park.) Result _____

11. The stars move 360° around the sky in 24 hours (1440 minutes). Over what fraction of a day did you observe? _____

Using the result from question 10, what fraction of a circle did the star traverse? How well do your answers agree?

If you could measure the angles very accurately (and also measure the east-west change in angle accurately) your results should be very close. A good result will differ by no more than 50%. Excellent results will differ by no more than 25%. Accurate observations like these were, until very recently, the most reliable method of timekeeping.

Do this part only if you are taking the course in the summer.

Date: _____   Location: _____

1. Find the bright star near the <u>eastern</u> horizon. Try to identify it on
   the star chart. What constellation is it in?_____

   Be sure that you can find this star again when you have finished the
   other parts of the lab.

   How many degrees is it above the horizon? _____ What time
   is it now? _____

2. The handle of the Big Dipper points to the brightest star in the
   constellation Bootes: Arcturus. Find Arcturus and measure the angle
   between it and the last star in the handle of the Big Dipper.

   _____

3. If you follow the line made by the handle of the Big Dipper and
   Arcturus, you will come to the brightest star in Virgo: Spica. How
   many degrees are there between Spica and Arcturus?_____

4. A little south and west of Spica are four closely spaced bright
   stars. They are the brightest stars in the constellation Corvus.
   Find this constellation and measure the angle between the two
   southernmost stars.   _____

5. What is the angle between the two pointer stars (Merak and Dubhe) in
   the Big Dipper?   _____

6. How many degrees is Spica from the nearest bright star in Corvus?

   _____

7. Find the stars in a constellation which appears on the star chart but
   is not mentioned in the introduction. Sketch the positions of all of
   the stars you can identify on the next page. Measure the angles
   between the three brightest stars in the constellation and show the
   angles on your sketch. Measure the angle between the brightest star
   and the horizon and indicate the angle and the time you made your
   measurement below the sketch. Be sure to indicate which star you
   used to make your horizon measurement.

8. Find the star you measured in question 1. How far above the horizon
   is it now? _____   What time is it now? _____

Finish the rest of the lab inside.

9. How many degrees has the star you identified in question 1 changed in altitude moved since you first observed it? _____
    In how many minutes? _____

10. The star not only moved higher in the sky, but also moved from east to west. You measured only the vertical component of the motion. To find the approximate total number of degrees the star moved, multiply your answer from question 9 by 1.4. (This number is valid only for the latitude of College Park.) Result _____

11. The stars move 360° around the sky in 24 hours (1440 minutes). Over what fraction of a day did you observe? _____

    Using the result from question 10, what fraction of a circle did the star traverse? How well do your answers agree?

    If you could measure the angles very accurately (and also measure the east-west change in angle accurately) your results should be very close. A good result will differ by no more than 50%. Excellent results will differ by no more than 25%. Accurate observations like these were, until very recently, the most reliable method of timekeeping.

Photo Credits

Pictures have been reproduced with permission from the following:

Cover photo: Palomar Observatory, California Institute of Technology

pp. 35, 36: NASA

p. 37: Lick Observatory

pp. 38, 39, 40: NASA

pp. 63, 64, 67: National Optical Astronomy Observatories

p. 66: University Corporation for Atmospheric Research/National Center for Atmospheric Research/National Science Foundation

p. 65: American Science and Engineering, Inc.

pp. 97, 98, 99, 100: Copyright R. A. Bell, University of Maryland

p. 118: J. C. Duncan, Mt. Wilson and Las Campanas Observatories, Carnegie Institution of Washington

p. 119: N. U. Mayall, Lick Observatory Photograph

pp. 124, 140, 150 and insert photos: Palomar Observatory, California Intitute of Technology

The use of the above photographs in this manual does not constitute an endorsement by the above organizations of this manual or any parts thereof.